ON THE MARGINS

The fen raft spiders of Redgrave and Lopham Fen

ON THE MARGINS

The fen raft spiders of
Redgrave and Lopham Fen

Helen Smith
Sheila Tilmouth

LANGFORD PRESS 2014

Langford Press, 10 New Road,
Langtoft, Peterborough PE6 9LE

www.langford-press.co.uk

Email sales@langford-press.co.uk

ISBN 978-1-904078-62-3

Design origination and typeset by
MRM Graphics Ltd, Winslow, Bucks
Printed in Spain

To our children and our children's children

Aerial view of Redgrave and Lopham Fen 1947 © English Heritage Archive

Contents

Preface

This book is our tribute to one of Britain's most beautiful and least common animals - which also happens to be a spider. It is not intended to be a monograph but rather a unique view of the species, bringing together our differing perspectives as artist and scientist.

We have told the spider's story largely in the context of Redgrave & Lopham Fen National Nature Reserve on the Norfolk/Suffolk border, where the first British record was reported in 1956. With only two more sites discovered in subsequent decades, Redgrave & Lopham Fen remains the most studied and also the most threatened of the British populations.

Helen – the scientist – over many years working on the conservation and ecology of fen raft spiders found that their elegance, maternal care, and struggle for survival against the odds, attracted consistent public fascination and media interest. This provided invaluable opportunities to use this species as a vehicle for telling wider stories about the plight of wetlands and their many unique and wonderful species. It also helped to promote interest in other spiders - perhaps our most misunderstood and misrepresented group of animals.

Sheila – the artist – first discovered the spiders and their story while surfing the internet. Immediately inspired by the potential for extending her artwork beyond the visual skin of representation, to encompass the cultural knowledge and scientific understanding of the subject, she contacted Helen with the proposal of an artist residency It offered exciting possibilities for taking the spiders' stories to new audiences.

Five years on, this book is the outcome of our collaboration. Sheila has developed an interpretive way of recording this spider's world. The twenty images that she created form the backbone of the book and are supported both by her photographs and commentary, and by Helen's text. Both art and science have the ability to transform our perceptions and the way we think. We wanted to bring the separateness of their focus together, to learn from and be inspired by these complementary views of the spider's world. Most of all, we wanted to share our privileged view of this rare and beautiful animal and to inspire others, not just to value it, but also to understand the importance of the complex network that binds a species into a wider ecosystem of plants, animals and the landscape on which they depend.

Foreword

I am thrilled to write a foreword for this superb book, not just because I can celebrate the grand architectural forms in Sheila Tilmouth's prints and collagraphs or the intimacy of her macro photography, and not simply so I can point out to you the precision and eloquence of Helen Smith's prose. I'm thrilled because I love work that is both aesthetically beautiful but also challenging.

Even the opening sentence to the preface gives you a measure of it. Helen Smith talks of 'one of Britain's most beautiful and least common animals.' Uncommon, certainly, given that the species only had three colonies until the author began her mission to reintroduce the beast to a string of other East Anglian locations. But a big hairy spider? Beautiful? I can already hear the arachnophobes start to shriek. But yes. Fen raft spiders are beautiful. That thorax and abdomen of deepest mahogany and then those clean cream stripes that frame and emphasise the exquisite body colour (while concealing its owner in its watery reflective home) – they give the raft spider an extraordinary appearance.

And like bitterns or swallowtails or fen orchids, raft spiders are beautiful in the way that they are supremely adapted to their wetland environment. We need to cherish them all. For too long British environmentalists have privileged a favourite few organisms and lost sight of, or merely implied the rest of nature – insects, fungi and spiders especially – in a kind of background haze. In truth what is important is the totality of it, the way it all knits together like an intricately patterned textile.

This book is much more provocative than simply asking us to love the unloved parts of nature. Its blend of art and science is itself a radical statement that says something profound about all wildlife and all environmentalism. A key personal concern over the last decade is the way we so often undersell the multiple roles that nature plays for the whole of society. A place like Redgrave and Lopham Fen, where Helen has studied raft spider ecology and where Sheila has painted the creatures, is not just a site of special scientific interest. It comprises a landscape and a complex community that affect the entire human spirit.

When artists and scientists work they are responding to the same sense of wonder. Their outcomes may be different, their routes may vary, but they journey to the same destination: to express some sense of the marvellous that inheres in all life. The sum of human experience is at stake when we lose natural landscapes, and when we lose spiders.

Just try to imagine the following without spiders: children's stories, traditional nursery rhymes, folk tales (especially West African), dusty cellars, B-grade horror movies, James Bond films, *The Lord of the Rings*, Louis XI of France, Scottish history, the very idea of persistence, Louise Bourgeois' glorious sculptures, Italian folk dance, American comic superheroes and finally that moment in late September when the dew gilds the silk across the sunlit lawn.

Spiders matter and this book makes the case perfectly.

Mark Cocker

Chapter 1
A SHORT HISTORY

Unbanded *Dolomedes plantarius*

1

On 24th June 1956 arachnologist Dr Eric Duffey made a remarkable discovery at Redgrave Fen in Suffolk. He found, and identified for the first time in Britain, one of Europe's largest and most spectacular species of spider. In his report of the discovery, he describes exploring a mosaic of different types of fen vegetation punctured by small ponds in the peaty soil. Examining the margins of one of these ponds he records 'I immediately noticed a large brown spider sitting on a reed stem. It was in fact the largest spider that I had ever seen in this country.[1] Duffey recognised the spider as belonging to the genus *Dolomedes*, the raft spiders. Although he expected it to be the raft spider then known from Britain, *Dolomedes fimbriatus*, this species is

Looking East over Little Fen

normally marked by conspicuous pale stripes along the sides of its body. Fortuitously, because this spider was plain brown, Duffey collected it. On closer examination, he identified it as a different species, the fen raft spider, *Dolomedes plantarius*. In fact only a small proportion of fen raft spiders are plain brown; most, like *Dolomedes fimbriatus*, are characterised by brilliant white or cream stripes along their flanks.

This book is the story of these large, striped and beautiful wetland spiders whose fate has been intimately linked with that of the site where they were first discovered; what is now Redgrave & Lopham Fen National Nature Reserve.

1 ON THE MARGINS
Technique: Oil painting
6.5"x 4" : 165mm x 100mm

At Redgrave and Lopham Fen where the fen raft spider, *Dolomedes plantarius*, was first discovered, dense vegetation of sedge and reed breaks at the margins of pools. From this position the spider can be observed, resting motionless, poised between the elements of air and water, its shape caught in shafts of light that penetrate through overarching leaves.

The water surface supports the spider but stretches under its weight, creating dimples of light around its outstretched legs. Sticklebacks pass by underwater dodging the tangle of vegetation that grows through the summer months.

The work is painted using a traditional technique of oil paint on a gesso surface. Gesso is a creamy mix of whiting and rabbit skin glue. Several coats of this mix are applied warm to a smooth, rigid wooden board. It hardens as it dries and can then be sanded to create a perfectly smooth surface: this makes it possible to apply paint in very fine detail, without the intrusion of a canvas weave.

Even the smallest painting takes weeks to complete: daily attention concentrated on a few square centimetres instils an intimate, meditative quality to the process, focusing all of the senses on the piece of work in hand.

This image uses a limited palette of five or six translucent colours, including Naples Yellow, Alizarin Crimson and Ultramarine. These are mixed to create layers of coloured glaze that allow tints of colour to penetrate through and emphasise the illusion of depth.

Fen pool

Fen raft spider

Stickleback

About raft spiders

In his still inspiring *World of Spiders*, published in 1958, W.S. Bristowe describes how raft spiders got their name; it was due to 'an error, published in the literature, which has been blindly copied ever since'. The error was a belief that the spiders could float down-stream on leaves bound together with silk. But perhaps the misnomer was not so misplaced. These spiders can both travel on floating debris - though not on deliberately constructed rafts - and can themselves behave like rafts, spreading their legs on the water surface and utilising the surface tension to support their weight.

Raft spiders belong to a family called the nursery web spiders, the Pisauridae. These are elegant spiders with cigar-shaped bodies and relatively long, robust legs. At rest on a solid substrate, the front two pairs of legs are characteristically held together, extended in front of the body, while the back two pairs extend together behind the body. These spiders have good eyesight and are active hunters. Webs have no part in their hunting armoury although, as their name suggests, they characteristically construct nursery webs – large, complex and conspicu-ous silk structures woven amongst plant stems - in which they rear their young. Their eggs and newly-hatched young are carried in a silk sac. In contrast to the wolf spiders (Lycosidae), which anchor their egg sacs with silk threads to their spinnerets (the tiny appendages from which silk emerges) at the rear of the abdomen, nursery web spiders hold their egg sacs beneath their bodies. They grasp them firmly in their mouthparts (chelicerae) and support them with their palps (the short, jointed, leg-like appendages either side of a spider's mouth), sometimes, additionally, attaching them with silk to their spinnerets.

World-wide, forty-eight genera of nursery-web spiders are recognised.[2] Among them the raft spiders, *Dolomedes*, span the globe. This genus is particularly diverse in south-east Asia while northern Europe has a very low diversity not only of *Dolomedes* but also of all Pisaurid spiders. Here there are only two genera; *Dolomedes*, the raft spiders, represented by two species -

Pisaura mirabilis: the nursery web spider

Dolomedes fimbriatus and *Dolomedes plantarius* - and *Pisaura*, represented only by the common nursery web spider *Pisaura mirabilis*.

Pisaura mirabilis is a common but often overlooked spider in Britain. Reaching to between ten and fifteen millimetres in body length, and varying greatly in colour and patterning, from subdued to striking, it is a species of sheltered grasslands and tall herbaceous vegetation. It is most likely to be noticed from late June onwards when females can be seen guarding their nursery webs. This spider is famous for its elaborate courtship, during which the male presents the female with a nuptial gift; a silk-wrapped fly safely engages her mouthparts during copulation, reducing any risk of suitor becoming prey.

The two European *Dolomedes* are much larger spiders and, like most of their genus, are found in close association with water. Often cited as being the largest British spiders, they have shorter legs than some house spiders (*Tegenaria* species) and bodies reaching similar lengths to the green-fanged *Segestria florentina*, but they nevertheless appear more massive and powerful. Bristowe's (1958) descriptions of them as 'splendid' and 'our grandest spider' cannot be bettered.

These species are very difficult to tell apart. Both have dark brown or black bodies decorated with brilliant white or cream lateral bands. Adult female *plantarius* have been recorded with body lengths of up to twenty-three millimetres whilst males are usually smaller; this so-called sexual dimorphism seems to be more pronounced in *fimbriatus* although, occasionally, *plantarius* males are only half the length of the females. To the experienced eye, the lateral stripes of *fimbriatus* are usually a little wider than in plantarius, the cardiac mark (a subtle, pointed blaze on the upper surface of the abdomen) more pronounced, and the body colours more variable. Spiders completely lacking lateral stripes are almost always *plantarius*. But, as with the majority of British spiders, definitive identification has traditionally had to rely on microscopic examination of the genitalia of adult specimens; the species cannot be distinguished at all for the greater part of their lives. Recent developments in genetic technology have provided the potential to identify spiders from their genetic fingerprints, finally allowing identification at any stage of development, though with costs and delays avoided by the traditional method.

A pirata species
with egg sac

2 DOLOMEDES THE MAGNIFICENT
Technique: Linocut
18"x 27.5" : 455mm x 700mm

This large and handsome specimen of *Dolomedes plantarius* has a deep russet brown coat with a creamy yellow band encircling the two sections of its body. It illustrates the spider in all its magnificence.

In its watery environment the spider inhabits a miniature world that we usually need to cross barriers of scale to observe. This piece of work enlarges the spider from its actual size of 7cms to 70cms magnifying the impact of its silhouette and representing in more detail the rich colour and surface texture covering its body.

Print making is a process of applying ink to a plate which is then offset onto paper either by hand or by using a specialist press to create an image. The plate can be made from a variety of materials, such as lino, or metal, or card. A relief print takes ink from the surface of the printing plate. An intaglio print takes ink from a groove on the printing plate.

Lino cut is a relief printing process. It involves cutting away parts of the lino to create a plate with exposed shapes that then has ink applied to it. Paper is placed onto the plate and pressed against it, either by being passed through a press or by being rubbed by hand. The areas of the plate that were cut away do not touch the paper, leaving just the exposed shapes to transfer ink to create the print.

Before cutting the lino the drawing for this image was refined many times ensuring the shape conveyed the powerful emblematic design of the spider.

Lino is softer and easier to work with when warm. The process can feel quite sculptural: the drawing is slowly revealed by cutting areas away, using the range of fine tools designed for this purpose. Mistakes cannot easily be rectified, so great care and concentration is needed at all times, in this case especially when shaping detailed identifiable areas such as eyes or spines.

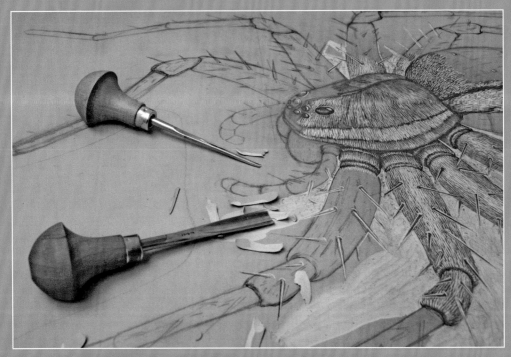

Cutting the lino plate

An 1830's Columbian Eagle Press

The first print

Raft spiders in Britain

Distribution

In the over half a century since the discovery of fen raft spiders at Redgrave & Lopham Fen, and despite greater awareness and recording activity, only two more populations have been discovered in Britain. The first, in 1988, was on the Pevensey Levels in East Sussex. This great low-lying tract of grazing marshes, much of it reclaimed from the sea in the Middle Ages, is intersected by a dense network of drainage ditches. As at Redgrave & Lopham Fen in 1956, it was a plain brown *Dolomedes* at the water's edge that attracted attention. Invertebrate specialist Peter Kirby recorded that 'it was quite unlike any *Dolomedes fimbriatus* which I had ever seen.[3]

Saw sedge beds at Redgrave and Lopham Fen

Subsequent confirmation of the specimen as *plantarius* triggered a large-scale survey of the Levels, by local teacher and enthusiastic arachnologist Evan Jones, three years later. This revealed an extensive population, dense in places though for the most part sparse and fragmented. Modern agriculture, particularly through impacts on the quantity and quality of water in the ditches, had taken a heavy toll on what must formerly have been a continuous population over an even greater area.

In 2003 Britain's third fen raft spider population came to light when wildlife photographer

Female reed bunting on saw sedge

Mike Clark was photographing lizards basking on a birch log on the edge of the Tennant Canal near Swansea in South Wales.[4] A movement further along the log diverted his attention – he recognised it as a large *Dolomedes* which, on expert examination, turned out to be a fen raft spider. Subsequent searches for the spiders in this area revealed further fragments of the population, linked by the canal and within the inaccessible and trembling wilderness of Crymlyn Bog, two-and-a-half kilometres to the west. In wet summers, they have also been found within the flooded mire of Pant-y-Sais National Nature Reserve adjoining the canal. Once again, the picture is of fragmented remains of what must have been a much more extensive and better connected population, broken up in this case by the impacts of industrialisation and urbanisation.

At first glance these three far-flung sites appear very different; a valley fen, a grazing marsh, and a canal and nearby bog. For the fen raft spiders, though, all offer the same critical elements. They have a long history of reliable, year-round supply of water that is rich in minerals but often relatively poor in nutrients. Under these conditions vegetation develops that is dominated by plants with an open structure, allowing the sun's warmth to reach the water below. It includes stiff-leaved, emergent species, such as saw sedge (*Cladium mariscus*), tussock sedge (*Carex paniculata*) and water soldier (*Stratiotes aloides*), that offer the complex and firm support preferred by the spiders' for construction of their nurseries, and it supports some of the richest communities of aquatic invertebrates in Britain. Parts of all of these sites are protected as National Nature Reserves.

Rarity

For many of our rarest species the geographical pattern and chronology of decline is well known and often influences their conservation status. But for British fen raft spiders, their 1956 discovery both denies them a history and raises many questions. How could such a large and striking species have been overlooked for so long and, without understanding its history, how can we now account for its bizarre distribution and extreme rarity? On close examination a catalogue of factors offers explanations. Those who have looked in vain for *plantarius* in its strongholds will testify to its cryptic nature. The jarring shock waves from an unwary boot-fall on water-logged peat will instantly banish many of the spiders into hiding below water. The camouflage of those remaining at the water margins is as frustrating for the would-be observer as for the would-be predator. Their body colours borrow from the subtle palette of fen muds and their bright stripes from the flashing curves of reflected sunlight where the waters' meniscus warps to accommodate emerging stems.

On wet fen sites the inaccessibility of the spider's habitat must have contributed further to the lack of records. Once found, confusion with the very similar *Dolomedes fimbriatus* was a major problem. First distinguished as a separate species by Clerk in 1757, many arachnologists continued to regard *plantarius* as a subspecies or variety of *fimbriatus* until well into the 20th century. Even after their specific identity became well accepted, the close similarity between the species continued to result in confusion, as did a largely unquestioning assumption that the only British species was fimbriatus.

Dig deeper and evidence emerges that all of these factors have been at play. Although Duffey's 1956 record was the first of a *Dolomedes* in the valley fens of the Norfolk /Suffolk border, the sedge cutters of the early 20th Century recalled seeing huge, striped spiders running out of their sedge bundles. By coincidence Mike Clark, discoverer of the Welsh population, had

3 REFLECTIONS
Technique: Etching
3.75" x 4" : 95mm x 100mm

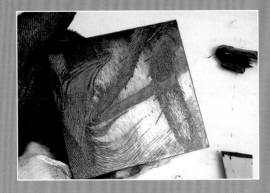

Inking and wiping the etching plate

Hidden in the broad uninterrupted landscape of Redgrave and Lopham Fen the fen raft spider is isolated in two small, disconnected populations. They are found near the water's edge and can be seen perched with front legs on the water surface and back legs resting on leaves, broken stalks or debris. Shifting sun and cloud in the East Anglian sky creates a canopy of moving light and shadow, helping to camouflage the distinctive shape of this spider.

Etching is an intaglio printing process. It involves creating marks or grooves on a metal plate, and smearing ink into these. The surface of the plate is wiped clean with scrim and then tissue paper to remove any ink residue. A fine art paper which has been dampened is placed on top of the plate, and together they are put through a printing press under considerable pressure. This forces the damp paper into the recessed marks where it will pick up the ink to create the printed image. The surface areas, where ink was removed, do not print.

To create the grooves a 'ground,' such as wax or acrylic, which is resistant to liquid, is first applied to the metal plate. Tools are then used to draw through the ground and expose the metal.

The back of the plate is sealed before the plate is plunged into a liquid chemical called the mordant. The mordant etches into the exposed metal, creating marks and grooves. The areas of the plate coated with ground remain unchanged. The plate is then cleaned, ready for printing.

This etching of reflections was created with a zinc plate. The jigsaw of shapes and curves thrown up by the meniscus, or water surface, makes it difficult to judge what is solid or shadow or reflection.

A wide East Anglian sky

Broken light falling on the margins

Reflections on the water surface

become familiar with raft spiders during childhood fishing trips on the Pevensey Levels in the 1950s. The local boys knew them as 'giant spiders'. Local naturalists certainly knew of the presence of *Dolomedes* spiders on the Levels by the early 1980s, but they were assumed to be *fimbriatus* until correctly identified by Peter Kirby. At the Welsh site, *Dolomedes* spiders had been discovered in Crymlyn Bog in 1978 by Friends of the Earth campaigner Andrew Lees, but were also assumed to be *fimbriatus*.

Although the current distribution of *Dolomedes plantarius* suggests that it must formerly have been widely distributed in Britain's formerly extensive lowland wetlands, none of the factors that contributed to its lack of verifiable history is likely to account for the paucity of new discoveries since 1956. Although some populations may still remain undiscovered, the discovery of only two more populations, despite increasing awareness and intensity of surveying, testifies to this species' genuine rarity. The fact that the known populations are all clearly fragmented and depleted by agricultural or urban development suggests that others may have succumbed to the unprecedented pressures on the countryside during this period.

This species' close dependence on unpolluted wetlands, and particularly on fen habitats, has made it especially vulnerable. Of nearly three-and-a-half thousand square kilometres of fen present in England in 1637, only a hundred remained by 1934; ninety percent of that has since been lost. In what appears likely to have been this species' former heartland - the intensively farmed English lowlands - fens are now less frequent, smaller and more isolated than elsewhere in Britain. Grazing marsh habitats have also been hard hit. In the Broads of Norfolk and Suffolk, thirty-seven percent were lost to arable agriculture between the early 1930s and mid-1980s. Of those that survived, almost all suffered erosion of the richness and rarity of their wildlife as agricultural methods intensified. The use of broad-leaved herbicides for the management of pasture and ditch margins, the enrichment of the water courses by artificial fertilisers, and the lowering water tables to allow expansion of arable agriculture, all took a heavy toll. Sadly, the picture is similar over much of continental Europe; although populations undoubtedly still await discovery or correct identification, the overwhelming picture is of fragmentation, isolation and extreme rarity. Often, they come to light only when threatened with destruction. Fen raft spiders are the only European spider species to be Red Listed by the IUCN (the International Union for Conservation of Nature and natural resources) in recognition of their vulnerability.[5]

Raft spiders at Redgrave and Lopham Fen

The ancient fen

At Redgrave & Lopham Fen, where they were first discovered, the fen raft spider's story has been a cliff-hanger. The Fen is an enigmatic place. It occupies an ancient and shallow chalk valley that cuts across East Anglia, marking the boundary between the counties of Norfolk and Suffolk. But rather than being occupied by one large river, this valley holds two small ones, meeting back-to-back either side of a low ridge of sand at the western end of the Fen. The sand is part of a three-dimensional patchwork of sediments that infill the old valley. The Ice Ages dumped their spoil of boulder clays, sands and gravels, to be sifted and sorted by water and frost. Late glacial lakes contributed layers of marl, a fine, shell-rich mud. Amongst and often blanketing these layers are more recent deposits of fen peat. On the eastern side of the sandy watershed, the river Waveney was born, not as a babbling brook, but a multi-headed seepage that oozed and filtered into rivulets and streams that worked their way laboriously through the fen sediments before uniting to flow east to the North Sea. On the western side of the ridge, the river Little Ouse emerged in similar fashion, cutting meandering tracks through now fragmented fens before joining the Great Ouse on its journey to the Wash.

It was the accumulation of peat in the valley, and the copious supply of water driving this process, that defined the area's identity as a fen. The water welled up through springs rising from the buried chalk and seeped into the fen margins where porous sand met impermeable clay. The buried valley's shallow gradient ensured perpetual waterlogging. Deprived of oxygen, the activities of the animals and micro-organisms that break down the remains of dead plants were inhibited. Seeds, bark, pollen grains and leaves were all pressed into an ever growing history of the Fen, written in deep black peat; radiocarbon dating reveals eight thousand years of peat accumulation. The micro-organisms able to exploit these oxygen-deprived conditions

Autumn morning

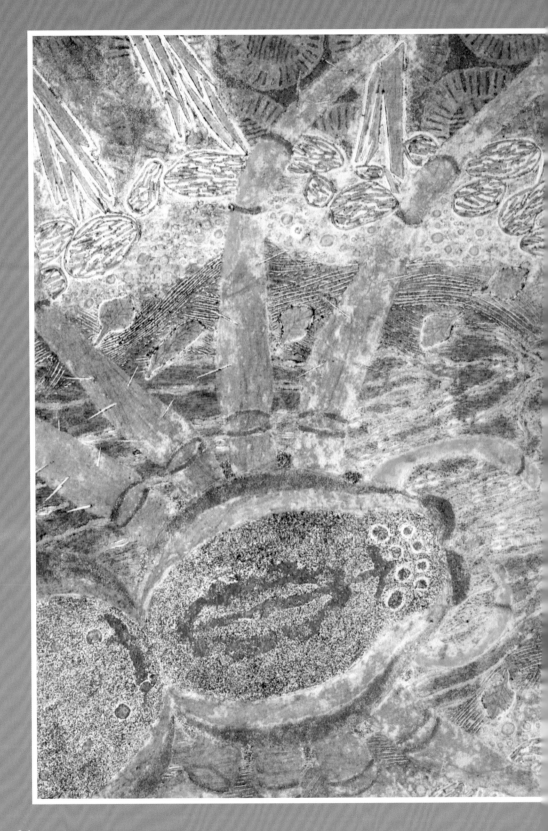

4 HERITAGE
Technique: Collagraph
15" x 15" : 380mm x 380mm

The earliest evidence of creatures resembling spiders, with eight legs and two body segments, is found from around 400 million years ago during the Devonian period when they were locked as fossils in a world that changed around them.

These ancient ancestors of the fen raft spider were evolving as the chalk that underpins the geology of Redgrave and Lopham Fen was forming. Now layered over with peat, clay, mud, sand and gravel, this stratum of chalk is important as it affects the Fen's ecology ensuring that it is irrigated with an alkaline supply of water.

Collagraph is an amalgamation of printmaking techniques combining elements of both relief and intaglio processes, usually using thin wood, metal or card for the printing plate.

A variety of materials such as cut out bits of sandpaper, tiny seeds or crumpled tissue are stuck onto the plate to make different surface textures and create edges or ridges that will hold ink in contrasting ways. Grooves and scratches can be added to the base plate to hold ink or be left clear in the printed image, according to how it is inked. Before ink is applied to the plate it must be sealed with a waterproof coating such as an acrylic varnish. Usually a press is used to make the print as the paper must be pushed into the intaglio element of the plate.

Thin plywood was used for this print and different materials glued onto it to symbolize some of the constituent parts of the Fen's geological heritage: leaf shapes of birch, willow and sedge represent peat; jagged shapes of card mimic glacial conditions and the shapes of boulders were made using coils of card. The advantage of using plywood rather than card for the base plate was that it could be cut with linocutting tools to draw the spider's spines and deep veins of the leaves.

In this image the spider is depicted locked in time, merged into a state of unity with the material substance of its surroundings: the colours are muted, the surfaces weathered and fused as one.

Early morning on the Fen

Section of collagraph plate

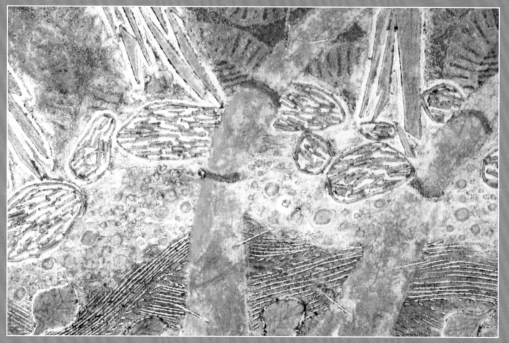

Section of print

Mist rising over Fen pool

added to the Fen's otherworldly atmosphere; in summer it became a place of strange colours and strange smells. Incongruous pink and purple bacteria stained the pool surfaces. The miasma of gasses rising from decaying plants could ignite spontaneously to haunt hot nights with ethereal willow-the-wisps, or mingle the sulphurous stench of black mud with the distinctive stale garlic odour of stoneworts (Chara species), bleached and stranded by receding water.

Mires such as Redgrave & Lopham Fen, formed on the sides of valleys and fed by water emerging from the ground, rich in minerals but poor in nutrients, are known as valley fens. The headwaters of the Waveney and Little Ouse held the largest expanse of them in all of lowland England. Where the chalky water emerged from the seepages and springs around the fen margins, it produced conditions ideal for a rich variety of low growing sedges and associated species, rare not only in Britain but also across Europe. As the water seeped further through the complex sand, peat and lake bed deposits, its chemistry changed, generating a small-scale patchwork of different associations of fen plants.[6] These ranged from chalky mires to acidic heaths, with the deeper peat supporting dense beds of aptly named saw sedge, now prosaically renamed great fen-sedge. Despite the exacting penalty extracted by its razor-edged leaves, this statuesque species has been exploited for centuries to waterproof thatched roof ridges. Unlike the brittle reed stems used to thatch the slopes of local cottage roofs, its long and flexible leaves could be bent over the ridge without breaking. The saw sedge beds of the East Anglian valley fens represent a significant proportion of this species' European population. Because of their rarity, these sedge beds, together with the low-growing sedge communities of the chalky mires, are listed as highest priority for conservation under the European Habitats Directive.

Saw sedge was not the only product supplied by the fens; the fen peat itself fuelled the fires of local cottages. Although peat digging had virtually ended by the mid-1920s when the railways brought cheap coal into the area, the Fen's surface is still moulded by its legacy. Dead-end footpaths from the villages still arrive via drag-ways at the sandy fen margins. Across the fen

Aerial view of Redgrave and Lopham Fen 1947

The late Albert Driver cutting peat in the 1970s
Photo David Orr

The late Albert Driver thatching a stack with Fen reed, sedge and rush in the 1970s
Photo David Orr

surface they branch into sinuous, raised baulks and narrow 'barrow ways' - arteries and capillaries spreading across the low-lying peatlands, still pocked by the remains of small turf ponds.

5 SEDGE
Technique: Collagraph
21" x 13" : 530mm x 335mm

Stout clumps of saw sedge cover the Fen creating banks of tall stems that outline the changing mood of the sky. The tough leaves are viciously serrated and lacerate bare skin. Their stiff upright structure creates architectural spaces around the edges of the open pools.

These pools are remnants of peat diggings that in East Anglia fuelled an agricultural economy for hundreds, if not thousands of years. The practice continued into the 20th Century as did sedge cutting for thatching. At the surface the pools appear small, but hidden below the water the sides fall away sharply creating deep pits. Alongside the spider, a diversity of small creatures has colonised this habitat.

The printing plate for this image is made from sedge leaves, dried out and flattened for the purpose over several weeks. They wilfully resisted adhering to the supporting card for this plate, and continued to exert a life of their own against attempts to bond them fast.

The leaves make parallel lines reaching for the sky, echoing their strength and presence in the Fen and reflecting the experience of a wall of grasses enclosing and embracing the pockets of openness created by the pools.

Many different applications of ink were applied to the plate and vertically wiped away. This creates a range of muted tints that express the changes that take place as sunlight fades or grey clouds lift during the moving cycle of days and months.

Serrated edge of sedge

Grasshopper

Sedge warbler

This pattern of human activity sustained and enhanced the Fen's natural richness. Removal of invading trees for firewood, grazing of stock on the higher ground, cutting of fen meadows for hay and of sedge and reed for thatch, and excavation of peat, all help to hold back the creeping natural progression from wetland to woodland. The necessity of digging new turf ponds ensured continuity of open water habitat, providing a reliable resort for aquatic and semi-aquatic species, including the fen raft spiders, in drier summers.

Desiccation

In retrospect, the excitement of the 1956 discovery of fen raft spiders at Redgrave & Lopham Fen was a bitter-sweet affair. The following year, changes began that came close to destroying this remarkable wetland and with it Britain's newly discovered fen raft spider population. An artesian bore-hole was sunk deep into the chalk aquifer underlying the fen sediments. Every day, over the next four decades, it supplied in the region of thirty-five thousand tonnes of clean water to local households. Remarkably, David Bellamy, then a PhD student studying the flora of Redgrave & Lopham Fen, is said to have noted that the chalk springs around the fen margins were turning on and off as the new pumps were tested. But in 1957 most of developments of this kind were unchallenged. It was to be more than forty years before developers would be required to assess environmental impacts as part of the planning process.

Despite increasing evidence that Redgrave & Lopham Fen was drying out, it was not until 1974 that investigations were instigated to explore the role of water abstraction in this process. The results were conclusive; although deep dredging of the river in the 1950s and '60s was an exacerbating factor, it was the artesian bore-hole that was primarily responsible for the Fen's demise. Although much of the fen surface was isolated from the underlying chalk by impermeable layers of boulder clay, occasional deep lenses of sand provided a connection. When the aquifer was full, water welled-up in these areas but as it was pumped dry they acted instead as plugholes, draining away the precious water from the fen surface. A site formerly fed by a copious supply of chalk water, low in nutrients, was now fed only by a combination of acidic rain and nutrient-rich water draining in from the surrounding agricultural catchment. As the surface peats dried and rotted, they themselves released centuries of stored nutrients, further adding to the problem.

As the Fen withered, recognition of its national and international importance grew. First designated as a Site of Special Scientific Interest in 1954, it had become Suffolk Wildlife Trust's first nature reserve in 1965. In 1991 it was recognised as an internationally important wetland under the RAMSAR convention and the following year it became a National Nature Reserve. Recognition of its importance in a European context followed, with designations as a potential Special Area of Conservation in 1995 and then as a Natura 2000 Site.

While these designations helped to bring the reserve's plight into the public eye, no amount of recognition could, in itself, halt the rapid loss of the Fen's precious wildlife. The growing list of species lost began with the rarest – those most dependent on the special wet fen conditions. Desiccation and nutrient enrichment of the fen surface brought impoverishment. A new set of species, better able to exploit the supply of nutrients, took control. Dominated by commoner vigorous grasses and increasingly by trees, they displaced the diminutive wetland rarities[7]. By the 1990s almost three quarters of the formerly open fen surface had been colonised by trees and bushes.

On the brink

As is so often the case, although the loss of plant species was well recorded, the Fen's rarest invertebrates tended to slip away unnoticed. Even for the raft spiders, by then the site's most celebrated inhabitants, records of their shrinking range are piecemeal. Eric Duffey, however, maintained a watch over them and, by the late 1960s, was becoming alarmed. The turf ponds on Redgrave Fen, one of four Fens that make-up the reserve complex and where he had first found the spiders, were often dry in summer and densely shaded by scrub. In an attempt to keep pace with the falling water table, new turf ponds were excavated – at first by volunteers with spades and then, in the 1970s and 1980s, on a much larger scale using excavators. Despite this, the situation continued to deteriorate, exacerbated by the increasing frequency of summer droughts.

Maintaining ponds for the spiders in the 1970s - *Photo David Orr*

Eric Duffey recalled searching in vain for the spider in the extreme drought of 1976 and 'wondered whether it had then become extinct'.[8] In the following years he found that numbers fluctuated in relation to the weather but that the 'general trend appeared to be one of population contraction and decline in numbers'. The recurrence of droughts in the late 1980s left the majority of the Fen's turf ponds dry in summer; clearly a very dangerous situation for an already fragile population. By 1990 the spiders probably occupied only around fifteen percent of their likely former range and were confined to two small and isolated pockets within the Fen complex.

Nominally protected since 1981 as one of only two UK spider species listed in the schedules of the Wildlife and the Countryside Act, in 1991 the spiders were recorded as Endangered on the UK's Red List of most vulnerable species.[9] In the same year the Government's conservation agency, the Nature Conservancy Council (now Natural England and other country agencies) began to fund more systematic monitoring of the spider population on Redgrave & Lopham Fen through its new Species Recovery Programme, instigated to reverse the declines of some of our most threatened species. Eric Duffey and his wife Rita undertook a detailed survey of the numbers of spiders on a selection of turf ponds within the two small and isolated areas of the Fen where the spiders were still found. Their conclusion, that the population had contracted to 'its smallest area and lowest numbers', was instrumental in triggering more radical conservation action. In August that year, a supply of water was piped across the Fen, from the pumping station adjacent to the bore-hole, to irrigate turf ponds in the two core spider areas. Although intended as an emergency intervention, the emergency, and irrigation of the ponds, continued throughout the 1990s. Although old ponds were deepened, new ones excavated and shading scrub removed to help the spiders, research evidence suggests that irrigation was the most critical measure in preventing their extinction.

The 1990s were a decade of watching and waiting: watching the remaining spider population through systematic monitoring, and waiting to see whether the conservation measures put in place would be sufficient to prevent its loss.[10] Waiting, too, for results from a long campaign to end the Fen's agony by re-locating the artesian bore hole; for all of the Fen's wildlife, any conservation measure short of restoring its sustaining chalk water supply could be no more than a holding operation.

A new chapter

The campaign was headed by the Suffolk Wildlife Trust, working in partnership with the water company and with government agencies. In 1993 this consortium bid successfully for European funding to help with the £3.2 million costs of relocating the bore hole and of re-creating

Clearing scrub in the present day

conditions that could allow the most important fen plant communities to re-establish. The ambition of this project cannot be overstated. As well as a massive and innovative wetland restoration project, it would be the first time that a major public water supply bore hole would be re-located solely on environmental grounds. The following year work started on clearing invading scrub and woodland from well over half of the Fen's surface. Desiccated and degraded surface peat was then stripped from a quarter of the surface to expose saturated layers of sufficiently low-fertility for fen vegetation eventually to re-establish. The Fen was fenced for the first time in its history, allowing the introduction of grazing stock to help maintain these conditions without resort to more traditional but expensive mowing. Polish Konik ponies, descendants of Europe's wild Tarpan horses and a now familiar conservation tool in wetland reserves, were first brought to the UK to graze this site.

Konik ponies

The immediate result of all this activity was that, by the late 1990s, much of the Fen was laid waste; the ancient human story imprinted on its surface erased. Bare tracts of peat, like clean sheets of paper, awaited the writing of a new chapter. In August 1999 the artesian bore-hole ceased operation and, during the following wet winter, water finally reclaimed the Fen, soothing its bareness and beginning the long process of healing.

The two areas where the spiders still occurred were spared the trauma of peat removal but attitudes to the spiders during the restoration operation became ambiguous. To many, these large, beautiful and desperately rare spiders had been torch bearers, shining a light on the Fen through its bleakest years. The European funding had clearly been awarded for restoration of conditions that would promote recovery of the Fen's internationally rare habitats. But negative newspaper headlines, equating such a large expenditure with such a small number of spiders, contributed to a view in some quarters that the spider story was no longer the most appropriate vehicle for public engagement with the Fen.

Winter flooding

THE EXPRESS: TUESDAY MAY 6 1997

'Don't worry dear, it's probably just looking for water'

Cookson cartoon. Credit Bernard Cookson, 6th May 1997 Express Newspapers,
British Cartoon Archive, University of Kent, www.cartoons.ac.uk

Bladderwort flower

During the following decade the Fen started - very gradually - to be re-born. Birds returned first, the bare scrapes drawing in migrating waders – wood sandpipers, greenshank and spotted redshank. The flooded peat surfaces rapidly greened with stoneworts – highly structured, fen-specialist algae that deposit tiny 'stones' of calcium carbonate over their surfaces. Although reed made a rapid advance on the new open waters, exploiting nutrients flushed from decaying peat, very slowly but surely more typical fen vegetation began to re-emerge. Reed gave way to saw sedge. The return of a low-nutrient regime was epitomised by carnivorous plants; once thought lost, delicate butterworts and sundews uncurled sticky leaves to supplement their meagre diet with insects. Bladderwort, dangerous below water and marked by carpets of

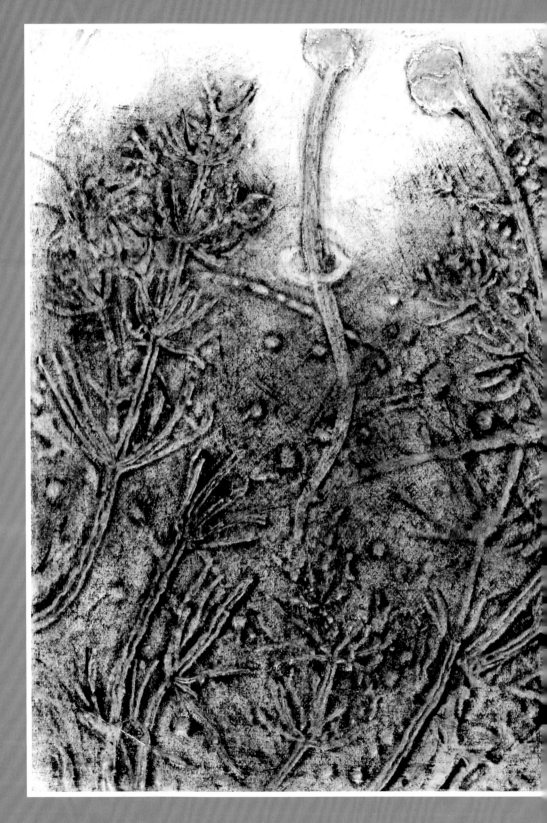

6 CHARA
Technique: Collagraph
7" x 6" : 180mm x 150mm

The alga Chara is predominant in many of the spider pools, greening the dark water in the spring with leaf-like structures of fresh delicate shoots reaching upwards towards the surface. Through the course of the summer it grows in abundance to fill the available space often rising above the level of the water surface as the Fen dries providing a welcome resting place for passing invertebrates.

These algae usually grow in neutral to alkaline water and are amongst the first species to colonize newly dug or reclaimed pools, as dormant spores can lie inactive for many years awaiting the return of favourable conditions before bursting into life. The large mass of the plant is readily populated by micro-organisms forming a rich food chain. In late summer the exposed tracts of Chara take on a ghost like, sculpted appearance as they dry chalky white due to encrusted deposits of calcium salts. These algae contribute to the distinctive odours that emanate from the pools at this time of year.

Although it is possible to reproduce approximately a collagraph print, no two are identical in the way that can be achieved with a linocut or line etching. This is due to the complex inking opportunities offered by the combined relief and intaglio properties of the plate. It will usually take several experimental attempts to obtain something near to the effect that the image was conceived around.

Orange fruiting body of Chara

Exposed Chara in peat pool

Ruddy darter on chalky white Chara skeletons

improbable canary yellow flowers above the new scrapes, harvested water fleas on an industrial scale with batteries of minute suction traps. Water rail and, more tentatively, snipe, returned as breeding species, haunting spring nights with the strangest of vocabularies. Bright dragonflies decorated sunny days, pursued by lightening hobbies. The Fen's riches were at least partially reclaimed.

But what of the spiders? Their undoubted potential for rapid increase, with each silk egg sac holding many hundreds of young, led to an expectation that numbers would respond rapidly to the returning water. But as the Fen re-wetted and flowered again it became clear that, for this species at least, recovery would not be so simple. Systematic annual monitoring of the size of the spider population failed to show any sign of a consistent or sustained increase.

By 2004, concern was once again translated into action. This time the most pressing need was not for provision of habitat improvements but for fundamental understanding of the spider's biology. Although for endangered birds, mammals and plants, research into their ecological requirements was a well-accepted pre-requisite for conservation management, invertebrates had tended to draw a short straw - 'best guess' approaches to their management were the norm. For the spiders, it was only the failure of best guesses to deliver population

South Lopham village sign

recovery at Redgrave & Lopham Fen that finally triggered funding for a research programme. By 2008, one Master's and two doctoral theses had started to unravel the spiders' secrets, from their day-to-day lives to the history of their isolation, recorded in their DNA. This new information now started to underpin radical developments in conservation action for the spiders, described in Chapter Four.

The spiders' tale had been linked for so long with that of the Fen that their failure to recover, and the new research this engendered, did not go unnoticed by the media. At the time it seemed to some an unwelcome caveat to the overwhelming success story of the Fen's recovery. But, once again, the spiders had an important story to tell the world. It was a moral tale. Although with huge expense and effort, we can now restore natural riches so easily and unthinkingly destroyed, the retrieval is never complete. The new is less rich, less varied, more commonplace, than the old. Some of the rarest, most specialised, most precious, never return while others, like the spiders, are changed by their experience and unexpectedly difficult to retrieve; perhaps the most salutary message to emerge from the new research was Master's student Andrew Holmes' finding that the spiders at Redgrave & Lopham Fen had lost significant amounts of their ancient ancestral genetic diversity, just in the last two decades. The spider that we were trying to save was not the spider it had once been – it was diminished, less able to adapt to face a rapidly changing world.

Threads

By 1992, the year of my first encounter with the spiders, Redgrave & Lopham Fen barely justified description as a wetland. Even the alarmed white flick of a moorhen's tail by Worby's drain, which fed the Fen with foul water from its agricultural hinterland, was a noteworthy record. Wellington boots were rarely a necessity. Although it still remained a wild and exciting place, fulfilling for me a long-held dream of living on the edge of a nature reserve, I was glad not to have known it in its heyday. Just as I avoid revisiting precious Mediterranean places– the sunlit slope where Hermann's tortoises crunched pink hawk's beard flowers or the marshes where black-winged stilts balanced in precarious display - for fear of finding only faceless concrete, I could not have lived with the loss.

With a background in plant ecology, spiders were for me an unthought-of future direction. My meeting with Eric and Rita Duffey on the Fen proved life-changing. Asked to complete their season's monitoring work on the spiders, I found myself holding a baby that could not be put down for fear of losing it forever.

I would like to be able say that I remember my first sighting of the spiders, but my early encounters in the company of experts were blighted by the secret fear, so often sensed amongst groups of bird watchers, of being found to be inadequate, incompetent at locating or identifying them. The memory has made me for ever sympathetic to those now trying to do the same. But over the twenty-two years since then, and many hundreds of spider sightings later, I can revisit many individual encounters with these remarkable animals and particularly with the mighty and intensely maternal adult females. One, watched through the nurturing of her first brood, found again with legs gathered in, dead by the water and still holding her second precious egg sac. Another, precarious with only four remaining legs, high in the sedge defending her brood. And standing deep in the mud of an ancient turf pond, seeing raft spider legs gripping fast around the underside of an arching saw sedge leaf high above my head, the body completely obscured by a large egg sac protruding over the leaf edges.

There have been years when every encounter seemed significant. During the droughts of the 1990s there were days of census work when the numbers of sightings were so low that I started to doubt my ability to find them. Even now the first sighting of familiar white stripes poised at the water margin in spring, heralded by chiffchaffs, yellow amongst insect-rich willow catkins, comes as a relief.

The early chapters of the fen raft spider's existence at Redgrave & Lopham Fen are unknown and unknowable. Since their discovery in 1956, their short history has often threatened to be very short indeed. But their fate had become so synonymous with that of the Fen that its recovery without them would have seemed a hollow victory. For me, privileged to share their extraordinary lives for the last two decades, it would also have seemed a personal failure.

Chapter 2

LIFE AT THE INTERFACE

Large red damselflies

Damselfly carnage

The fen raft spider is elegant, adapted and at home in two very different worlds, air and water. Many animals make this transition but few with such complete mastery. Most are perfected in one but clumsy, vulnerable, fly-by-night in the other.

But this spider's real mastery is neither of air nor water but of their meeting place, the water surface or meniscus, where the force of surface tension pulls water molecules into a tight, elastic skin. For many animals, terrestrial and aquatic, this is a place of necessity, a busy thoroughfare. A rite of passage for emerging dragonflies, dragging their soft new bodies, still darkly encased in the old, from water to air for the final brilliant metamorphosis. A place to be risked again as gravid females, depositing their eggs. A necessary step for amphibians on their yearly migrations, from the leaving of tiny froglets, deprived of gills and tail, to the spring return of adults to chorus, mate and spawn. An imperative journey too for terrestrial species needing water to quench thirst and for those aquatic species that need air to breathe, their evolutionary return to water incomplete.

For all these visitors - the thirst quenchers, the egg layers and the travellers – the meniscus is a place of danger. For the unwary lightweight, caught off balance in a sudden breeze, it is a sticky trap, a place of temporary struggle against the adhesive energy of the surface. For others, the greatest danger lies in the hunters, those making their living by exploiting the necessary passage through and to this place. Like leopards and crocodiles at the water hole, they wait at the water margin and beneath the water. Sharp-eyed sedge warblers bracketed on reed stems seize their opportunity to take visiting dragonflies. Cobwebs, brightly beaded in early morning dew but invisible in the heat of the day, are slung like fishing nets between emergent plant stems to exploit the incessant insect traffic. Grass snakes criss-cross the water, yellow-necked

in slow waves of stealth, to grab the unwary frog. But the hunted is also the hunter. Frogs line the water margin, sticky-tongued opportunists waiting for a slug seeking essential moisture or a passing wolf spider, busy on its own hunting mission. Just below the water surface the shadowy profiles of greater water boatmen – the backswimmers – ripple slowly, shark-like. Fish too – sticklebacks in the fen pools and fierce little jack pike in the ditches and drains – wait in the shadows for the touch of prey to dent the water surface above.

Masters of the meniscus

But amongst the many species that must pass through the meniscus, and those that exploit their passage, very few make it their own domain. Of these, largely unseen below water, some walk beneath it, like flies on a ceiling. Others hang from it. The larvae of mosquitos, bee-like chameleon flies and hoverflies, hang suspended by breathing tubes or by bristly tufts that funnel air to their spiracles. More conspicuous masters of the meniscus are those living on the surface, most of them bugs (Hemiptera) whose relatives live almost exclusively either on land or underwater. Pond skaters, or water striders (*Gerris* species), are the commonest of these meniscus dwellers. Protected from wetting by a fine, dense pile of water repellent hairs on their legs and undersides, they skim across the surface, their flimsy weight spread between six legs which dimple the meniscus, projecting concentric circles of shadows onto the mud below.

While pond skaters constantly patrol the meniscus for its dead and dying insect victims and for friend or foe, they neither enter the water nor emerge onto land, save for occasional furtive flights in search of new water bodies. Only when they first emerge from submerged eggs and swim awkwardly upwards to puncture the waiting surface, or as females returning to penetrate

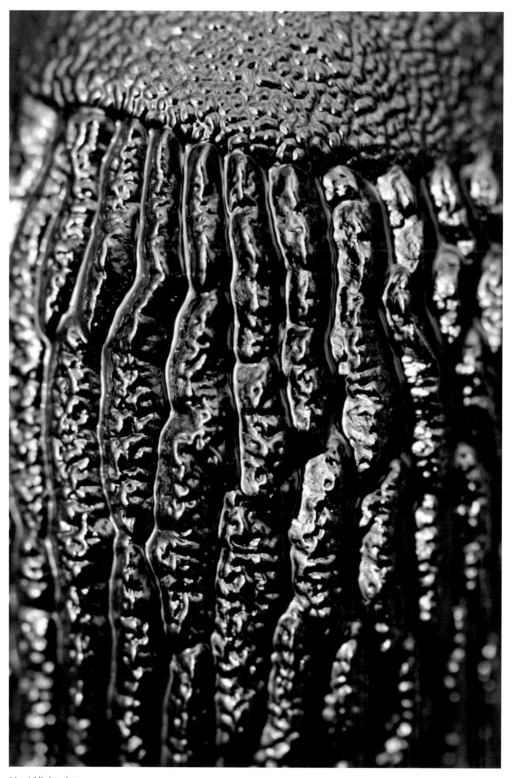

Liquid light: slug

Liquid light: reed reflections

7 HAIRY SPIDER
Technique: Etching
12" x 13.3" : 305mm x 340mm

After dipping rapidly underwater this male spider slowly emerged several minutes later, resurfacing and then climbing cautiously, with barely perceptible movements, above the level of the water to rest on an adjacent stem. This exposed a clear view of the underside of his complex anatomy which was, however, completely dry. Fine hairs that cover the spider's entire body form a barrier of air when it is submerged. This gives a ghostly silver sheen to its appearance when viewed under the water.

The spider has two sections to its body – a fused head and thorax called a cephalothorax and an abdomen that is joined to this. The cephalothorax is the hub for most of the spider's moving parts and its multi-jointed legs with thick spines are clearly defined in this ventral view of its body.

The spider offered a dramatic pose outlined above the water surface and showed in fine detail the structure and surface texture of its body. Articulations of the legs, together with the water repellent nature of the hairs that cover the body, are emblematic features of this semi aquatic species.

Detail of the radial structure of legs and body was carefully referenced for accuracy. It is a wonderful example of remarkable functional biological engineering, creating a stunning visual design. Etching was chosen as the most suitable print medium for this image as it lends itself to fine detailed drawing.

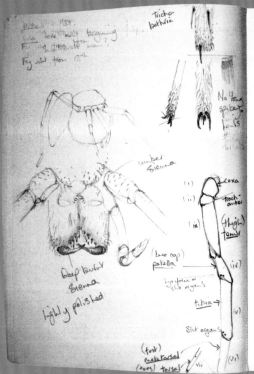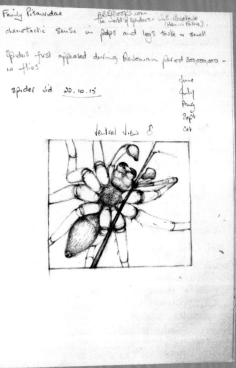

Sketching the anatomy

Refining the print

Underwater spider

Emerging spider

View over the meniscus

it again to deposit their eggs, do they enter the water. The raft spider alone not only masters the meniscus but can also turn its back on it to join the terrestrial world, or shatter it to hide or hunt below water.

A dense covering of highly water repellent hairs on the spiders' legs and body traps a thin cushion of air - the plastron – that helps them exploit the meniscus without risk of becoming wet. Like the pond skaters, their bodies deform the surface like a trampoline, the upward pressure of the dimples created by their feet, supporting their weight. The plastron also allows the

Stretching the meniscus

Silver underwater spider

spiders to breathe and to survive for extended periods underwater. The effort clearly required for the spiders to break the surface tension and enter the truly aquatic world below is rewarded by the foiling of terrestrial predators and the access to aquatic prey. Shadowy below water, glinting silver in its encasing sheath of air, the spider waits, holding fast to a stem or root, or swims between underwater perches, constantly fighting the air-bubble buoyancy that drags it upwards. Amongst the spiders, only the water spider, *Argyroneta aquatica*, master-builder of silk diving bells, is more fully adapted to life below the surface.

Getting around

On land the raft spiders adopt the stepping pattern typical of terrestrial spiders but the meniscus demands new, specialised gaits. Rowing, with the body in contact with the meniscus and the legs acting as oars, is the commonest means of travel. In the sudden urgency of prey pursuit or predator evasion, the gait changes dramatically. Galloping across the surface, alternating powerful leaps and free flights, the spiders can travel at up to seventy-five centimetres a

Sailing - *Photo: Arthur Rivett*

second, five times faster than by rowing.[1] In light winds, sailing is a much less demanding means of travel. Effective sails can be erected by raising one or more legs, or by standing on tip-toe, raising the body to catch the breeze. For all sailors, the wind must be carefully judged – too strong and it risks fatal instability. For the spiders, sailing brings the added risk of inability to control direction, but light breezes may offer the chance of a rapid and energetically inexpensive return to shore. When accidentally landing on a water surface, their cousins, the terrestrial and rather less waterproof nursery web spiders (*Pisaura mirabilis*), sail even more readily, seemingly recklessly, almost certainly in the hope of an aerial re-launch or a quick return to a safer, drier perch.

Only in extreme youth and old age does water seem to be a threat to the spiders. Leaving the safety net of their nursery webs when only a few days old, they head not for water but for land; the risk of their tiny bodies sticking to a rough water surface is too great. But whilst sailing and rowing are ill-advised for these tiny spiderlings, flying is a real possibility. In a bid to find pastures new, many species of small spiders, and the tiny young of some larger species, travel substantial distances by so-called 'ballooning'. Climbing to a high vantage point, they stretch their legs and stand on tiptoe. Tilting skyward their spinnerets at the end of the abdomen, they cast out a hopeful line of silk to the breeze. Drawn upwards by air currents, the pull on the silk is eventually sufficient to lift the spider into the air. Fen raft spiderlings show elements of ballooning behaviour just at the age when they leave the safety of the nursery, but research and field observations show that they are generally unadventurous. Streaming upward to reach leaf-tip vantage points, they then seem constrained to wait until their silk lines catch on near-by stems before tightrope-walking these modest canyons – so-called 'rigging' behaviour.

Rigging spiderlings

Flying at the whim of the wind, with no capacity to steer or choose a landing site, risks all. For isolated populations of rare species with specialised habitat requirements, this risk is likely to be too great. The chance of sufficient balloonists landing together in suitable habitat to establish a new population is so vanishingly low that natural selection removes this behaviour from their repertoire.

For adults too, nearing the ends of their lives, water is no longer their element. The meticulous grooming needed to maintain their essential waterproofing seems to be too demanding. Instead of perching on their backs and legs in tight spheres, water droplets, cling, spread and threaten drowning.

Although the many modes of locomotion available to the spiders offer an active life-style, for the most part this is a stay-at-home animal which uses as its base small stretches of water margin. Here it can often be seen day-after-day, stretched along a stem soaking up sunshine, or perched, poised and hunting. The typical hunting posture is head down, with back legs on plant stems emerging from the water, and front legs – anything from one to three pairs - resting on the meniscus. Water is an excellent conductor of vibrations and the raft spider's legs have excellent vibration receptors. An array of specialised hairs, the trichobothria, form the front line in a sophisticated early warning system, able to distinguish location, direction, frequency and strength of minute movements of air and water. In contact with water they give early warning of the approach of prey at or below the water surface, or of the struggles of terrestrial

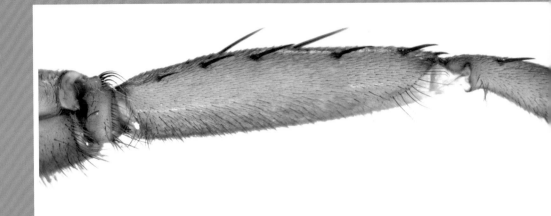

8 THE LEG
Technique: Digital image
11.7" x 16.5" : 300mm x 420mm

The design of a spider's leg is inspiring, being both aesthetically pleasing and uniquely practical with many different specialist features. Each leg consists of seven segments and each can move independently. The leg attaches to the body by a segment called the coxa. Attaching to this is a short, ball like segment called the trochanter. The third segment of the leg, the femur, is a large finely shaped cylinder ending in an articulated knee-like segment, the patella. Following from this are three long tapering segments; the tibia, metatarsus and tarsus.

The tip of the tarsus ends in three claws, two of which are toothed. These aid climbing on vegetation and are able to hook silk. Galloping spiders rushing across the water surface in pursuit of prey use the hooks as their braking mechanism, grasping their silk draglines to bring them to an abrupt halt when they reach their goal. The legs can compress tightly against the body or extend outwards into a radial display. Muscular action is used to pull the legs in but hydraulic pressure is used as well to extend them out again. The greatest degree of movement is through the coxa–trochanter joint while the joint between the tarsus and metatarsus is passive. This is reflected in the gait of the spider.

With weight distributed through eight legs, the spider's footfall on the water surface is light. It rests with anything from the tip of a toe to most parts of the leg on water. When moving it uses its legs to row, or gallop, over the water surface but with a firmer footing it moves by bringing each leg forward in turn.

Vital sections on the legs house comb-like structures for grooming mouth and body parts whist grooming also maintains the crucial waterproofing qualities of its fine hair covering.

Important sensory apparatus is found on the spider's leg. Slits register external pressure and vibrations against the rigid outer cuticle that encloses the spider's body – the exoskeleton. Chemosensory hairs are sensitive to taste. The most delicate hairs, called trichobothria, extend from the leg at regular intervals. They are able to discern a light breath of air or a vibration signalling an alert to the spider and convey essential information about movements of potential prey or predators when the spider is hunting.

The exoskeleton of the spider's leg glows amber in the strong illumination needed to photograph this tiny fragment, which was taken from a discarded skin shed during a moult. The same effect is seen in the field when a spider rests in a shaft of sunlight.

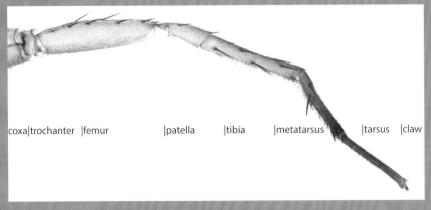

coxa|trochanter |femur |patella |tibia |metatarsus |tarsus |claw

Naming of segments

Grooming

Grooming

Tarsal claw and fine, protruding trichobothrial hairs

Fen raft spider; eight simple eyes

Hornet; two compound eyes and, three simple eyes

Dragonfly; two compound eyes

insects sticking to the meniscus. They warn too of the approach of friend or foe, distinguishing easily between the vibration signatures of approaching prey, predators and potential mates. In air they warn of threats – the rush of air as the lethal, sticky tongue of a frog catapults towards them, eliciting an evasive vertical leap from their perch or even from the water surface.

Perhaps surprisingly for a species with eight eyes, bestowing an almost a 360° range of vision, sight seems to play a relatively minor role the raft spider's ability to detect its prey and predators. The trichobothria alone are sufficient to elicit the evasion response to predators. Vibration rules the senses in this watery world.

The sense of taste - an ability to detect chemical signals - is also mediated through specialised hairs on the legs and palps. Taste may play a role in prey recognition but is better known as an important adjunct to vibratory signals in the recognition of potential mates (Chapter 3).

Prey and Predators

As predators, fen raft spiders are catholic, formidable and effective, on land, at the meniscus and under water. They take prey often much larger than themselves, including species infamous for their voracity, such as the massive and fierce larvae of hawker dragonflies (*Aeshna* species) and great diving beetles (*Dytiscus marginalis*). Neither are they restricted to invertebrates. The inclusion of both amphibians and fish in the raft spider diet plays large in their

Hunting posture

perennial ability to attract media attention. In practice, as ambush hunters, they probably take prey as they encounter it – small or large, common or rare. Most of their diet is comprised of common species of the water and water margin; water beetles and boatmen, dragonfly larvae, semi-aquatic *Pirata* spiders, and tadpoles in season. Hunting from underwater perches, they may particularly target concentrations of water boatmen or seek out sticklebacks. Hapless terrestrial invertebrates, ill-equipped for escaping accidental landings on the water surface, provide much easier meals.

Although fish-eating is frequently recorded in some other *Dolomedes* species in the wild and there are many reports of it in fen raft spiders, almost all of these involve animals in captivity.[2] Most of the verified observations of this behaviour in the wild are of the fen raft spiders at Redgrave & Lopham Fen. Small fish are most likely to fall victim when oxygen levels in the water drop and they are struggling for life. In the Fen turf ponds in high summer, the shallow layer of warm sulphurous water, filled by decaying stoneworts and surmounting a deep plume of back mud, is a challenging environment for the ten spined sticklebacks (*Pungitius pungitius*) that live there. The waiting raft spiders are perfectly placed to exploit their distress as they are forced towards the surface, gasping for the ever diminishing supply of oxygen.

Watching raft spiders hunt is a tedious pursuit. They catch when they are hungry, when the moment is right. When it comes, it is brief; a lurch or leap, a split second's struggle of tangled limbs and turning bodies - and then stillness. Using its legs to entrap the prey, often flipping with ease onto its back in the struggle, the spider's sharp fangs exude a venom that brings rapid paralysis. The meal can then be consumed in a place of safety and at leisure, leaving

Meal left-overs

9 THE CATCH
Technique: Hand tinted linocut
10.2" x 6.3" : 260mm x 160mm

The fen raft spider is an opportunistic, patient hunter which waits for prey to pass close by before lunging out to rapidly grasp its target. Immobilizing poison is quickly injected into the victim, swiftly calming its bitter struggle. Palps - the two leg-like structures found on either side of the spiders mouthparts - are important sensory appendages in the catch and assist with holding and manipulating the caught creature whilst fangs puncture its defences releasing venom from a specialist gland.

Spiders' eyes register patterns of changing light, but the sensory apparatus employed during the catch is found mainly on their legs and palps. Here the army of specialist hairs and spines register taste, vibrations and external movements.

The classic pose for a hunting raft spider is with front legs resting on the water surface and others securely positioned on vegetation or debris. In this way, the tiny vibrations of a pond skater or passing fish are registered through the tight drum of the water's meniscus. The spider can move at great speed and the embrace of legs, palps and jaws creates a robust cage making it difficult for even a much larger animal to escape.

After the catch has been paralysed it is infused with digestive juices that break it down allowing the spider to ingest its prey as liquid. The many hairs and comb like structures in and around the spider's mouthparts ensure that solid particles are separated out and discarded. The spider has no ability to grind down solids and only liquid, pre-digested food is drawn into its large capacity sucking stomach.

The image has an obvious drama as the spider grips the fish securely in its jaws. It incorporates all the elements that play a part in the catch: head, eyes, jaws, legs, water, light and stems. Keeping the shapes tightly composed within the outline of the printing plate, with the features explicitly enlarged, expressed the trapped claustrophobia of that moment.

The strength of linocut as a print medium is in the design of how light and dark play out in the composition. The eye of the fish with ripples of concentric circles around the pupil is a powerful focal point for juxtaposing the other elements of the image, whilst the texture of scales, hairs and ripples are visually interesting to contrast.

Muted colours of the fish, spider and water have been added by hand to the black and white print to give more depth to the finished image.

Catching a waterboatman

Catching a stickleback

Mouthparts – chelicerae above, maxillae below

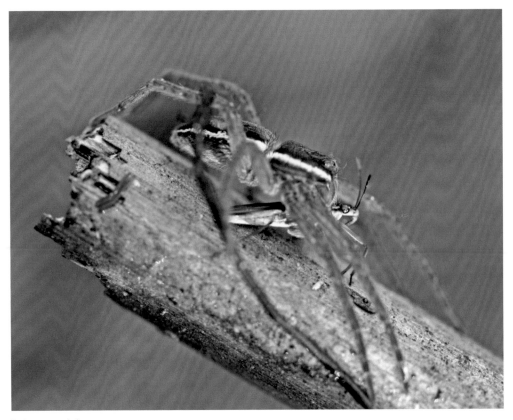

Eating a pond skater

behind a temporary floating memorial of chitinous[3] elytra (the hardened forewings of beetles) or rainbow wings.

Avoiding being eaten is as imperative as the need to eat. The spiders appear to have two main lines of defence – avoiding detection and evading attack. Avoidance of detection is conferred by their camouflage. The muted browns and blacks of the mud, disrupted by criss-crossing stems that reflect bright arcs of sunlight where they emerge from the water, are all mirrored in the spider's body colours and outline. Although there is much variation in the spiders' body colour, in the width and colour of their lateral bands, and in the occasional presence of spots on the abdomen, one difference in colouring stands out from all others. In all British fen raft spider populations up to thirty percent of the animals lack bright lateral bands. By crossing unmated adults with and without lateral bands, 'A Level' student Alice Baillie showed that the presence or absence of the bands is determined by a single gene with dominant and recessive forms. If a spider inherits the dominant form of the gene from either or both of its parents, it will be banded. Only if it inherits the recessive form of the gene from both parents will it be plain.

The main plank of the spider's second line of defence, the trichobothrial sensory system and the escape reactions that it triggers – jumping vertically, leaping from a perch or vanishing underwater – is effective in foiling many predatory attacks from both above and below. Research suggests though that some larger fish species can present a real threat; the spider's

Reflected sky

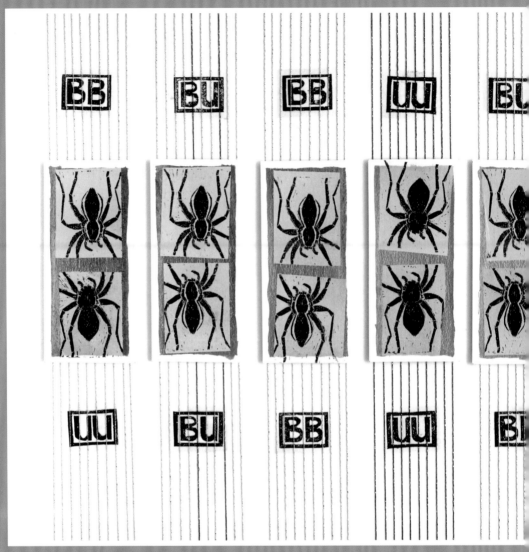

10 GENETICS
Technique: Mixed medium
13.5" x 25" : 345mm x 635mm

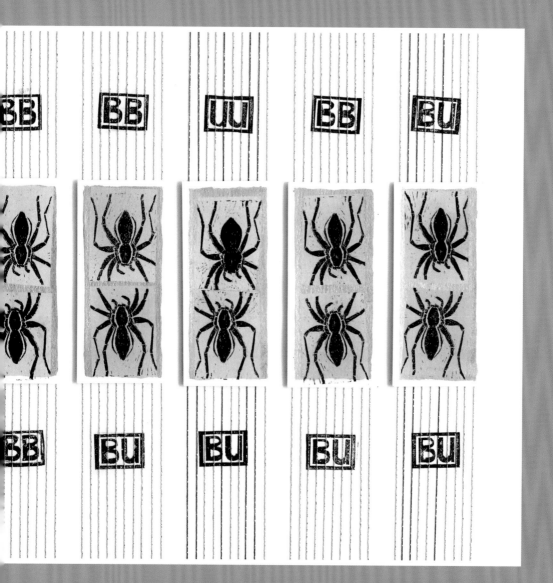

The genetic profile of an animal is a unique fingerprint. This coding is a key that underpins every characteristic of every species.

The presence or absence of a distinctive light banding on the body of the fen raft spider is a clear variation in the species that has been analysed for its occurrence on the Fen, with unbanded spiders accounting for up to 30% of the natural population.

Breeding experiments reveal a pattern of inheritance of the bands that is typical of a characteristic defined by two forms of a single gene, one of which is recessive and the other dominant. The recessive form gives rise to unbanded spiders and the dominant form to banded spiders. The offspring inherit one copy of the gene from each parent. If they inherit two recessive forms of the gene they will be unbanded. If they inherit one or no copies of the recessive form they will be banded.

This form of inheritance governs many characteristics of plants and animals and is called "Mendelian" after Gregor Mendel who first discovered it in the 19th Century from experiments on the inheritance of characteristics in peas.

This piece of work highlights the core role that genetics plays in defining the characteristics of individuals from one another. Illustrating the genetic principles that govern variation in the spiders' banding reflects back to the discovery in 1956 of Redgrave and Lopham Fen's *Dolomedes plantarius* population. Dr Eric Duffey, an arachnologist, seeing an unbanded spider atypical of the usually striped *Dolomedes fimbriatus*, collected it and on microscopic examination discovered that it was in fact *Dolomedes plantarius*.

The distinctive banding is a powerful symbol for this spider. I used gold leaf, applied onto card in the image, to convey the associations of preciousness and rarity. The shape of the spider in this case is cut quite crudely from a woodblock and hand- printed onto translucent Japanese paper to be glued over the gold leaf.

The parent spiders are paired and the letters B or U represents the banded and unbanded forms of the gene. Each parent has two copies of the gene and will pass one of these to their offspring through their egg or sperm cells. The strands of black and gold thread represent the proportion of banded and unbanded spiderlings expected in the broods of parents carrying different forms of the gene. The diagram illustrates this pattern of inheritance.

The line of spiders spread out for scrutiny in this image mirrors the methodical and systematic nature of scientific research in its endeavours to obtain a proven body of knowledge about these creatures.

Section of image

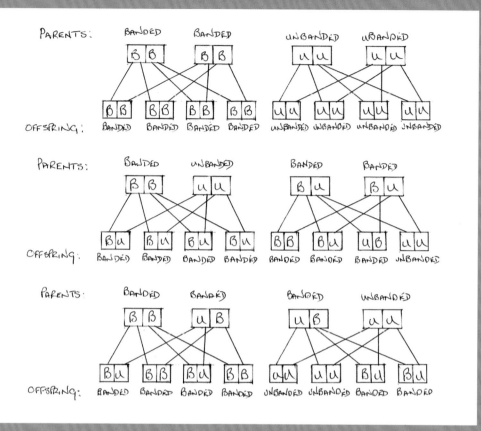

The pattern of inheritance for the banding gene

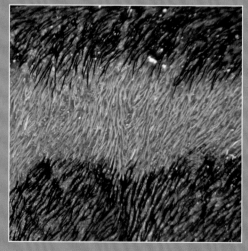

Skin deep: lateral banding in body hair and the shaved skin beneath i & ii

evasive jump from the water surface, though completed within milliseconds, is unlikely to be adequate to foil determined submarine attacks.[4]

If the trichobothrial system fails, a more radical solution can sometimes be used to get out of trouble. Spiders share with many other invertebrates a primitive ability to shed limbs grabbed by a would-be predator, leaving the spider free to escape. This so called autotomy appears to involve minimal trauma – the jointed limbs break easily at their weakest joint, close to the body. Although such limb loss is likely to result in some loss of speed, or of prowess in courtship and mating (Chapter 3), this is a small price to pay for escaping death.

Although the list of potential predators of fen raft spiders is long, there are few records of predation in the wild. In Britain, spiders have occasionally been seen disappearing into the wide gape of a frog. A neatly laid-out set of eight legs found floating on a Pevensey Levels ditch suggests an attack by a warbler; reed and sedge warblers are abundant both here and on Redgrave & Lopham Fen. Although there are no direct observations of warblers taking fen raft spiders, a photograph encountered on display in an Italian National Park shows a fan-tailed warbler holding a large raft spider firmly in its beak. Standing tall, with long reach, keen eye and stealthy foot fall, herons are obvious predatory candidates but to date without direct evidence in Britain. They are implicated though by repute - a record of thirty-two raft spiders (*Dolomedes triton*) in the stomach of a Little Blue Heron in the USA[5] demonstrates their lethal potential. But the large numbers of often very conspicuous female fen raft spiders found guarding their nurseries on water soldier rosettes carpeting grazing marsh ditches, suggest that at least for large spiders, the dangers of exposure may not be great. The dense web may itself provide efficient protection – guarding mothers often sit under its awning and vulnerable females carrying cumbersome egg sacs may also find shelter there. The bodies of females that die in their nurseries, worn out by the effort of breeding, often remain intact, unpredated.

In addition to the ever-present threat from predators, parasites and diseases must take their toll but our knowledge of these, even in common spiders, is at best scant. A variety of hymenopteran insects are known to predate spiders. Amongst these a family of solitary wasps, the Pompilids, are spider specialists. Although there are no records of Pompilids taking fen raft spiders, a North American species of Pompilid wasp has been observed successfully pursuing huge adult *Dolomedes triton* into their underwater boltholes. The wasp paralyses its spider victim with a venomous sting before dragging the helpless body back its nest. There, it lays a single egg on the spider's abdomen. The hatching larva feeds on the still living spider, its sex bizarrely determined by the spider's size. If the spider is small, the egg remains unfertilised and produces a male wasp; if larger, the egg is fertilised and produces a female wasp. In addition to the Pompilids, an array of other parasitic wasps and flies are potential threats to the spiders but danger may also come in the more insidious, little studied, form of microorganisms and wetland fungi that envelop, invade and kill.

For the youngest spiders the world is an even more threatening place and the list of potential predators much longer, though even more difficult to observe and quantify, than for their formidable parents. Leaving the relative safety of the nursery at only a couple of millimetres in length, the spiderlings must fall prey to many species, amongst which other spiders, as well as birds, are top candidates. Even minute, translucent and immature *Clubionid* spiders, fellow inhabitants of the damp fen vegetation, will predate fen raft spiderlings. Many species that will eventually become prey to those raft spiders that survive to become large and powerful, now pose a predatory threat - the food web that links these species is both complex and dynamic.

Common blue butterfly at rest

11 FOODCHAIN
Technique: Linocut
10.6" x 14.5":
270mm x 370mm

Late spring sees an awakening of life on the fen as temperatures rise and days lengthen, unlocking animals and plants from winter dormancy. Eager birdcall fills the air as partnerships are made or lost and the water margins are abuzz – the major highway for a host of bugs including emerging damselflies and pond skaters. Leaves unfurl and flowers offer up their nectar to a throng of multifarious insects. The pools teem with life – vertebrate and invertebrate, amphibian and reptilian, aquatic and terrestrial – all focussed on survival, with an overwhelming compulsion to feed and reproduce.

The foodchain on the fen is complex: a species may be predator or victim at different stages of its life cycle, as egg, larva or adult. In this way a spider might eat a tadpole but, in turn, be dinner for a frog. The complex interdependence of small wetland creatures supported by this habitat allows a great diversity of them to co-exist.

In a linocut there is a direct relationship with the image as you guide tools around the contour of a shape and pare away superfluous detail. Often the sketches drawn beforehand are inconclusive and the final positioning of a form is left to fall into place organically as the design evolves.

In this image the horizontal banding running through the composition symbolises the range of elements associated with the needs of creatures in this foodchain: solid ground or water, a platform, places of safety, an horizon or stratification, ripples…the list goes on, but left ambiguous, it can be any of these things, and individuals from the great variety of fen animals find sympathetic spaces within the scheme.

Tonal variation is achieved in a linocut by altering how much is cut away changing the proportion of white paper to inked surface. It can be controlled and regulated methodically, grading the plate from black to white or as in this case, intuitively creating a visually engaging relationship between areas of light and dark that will lead the eye across the whole image or to a point of focus.

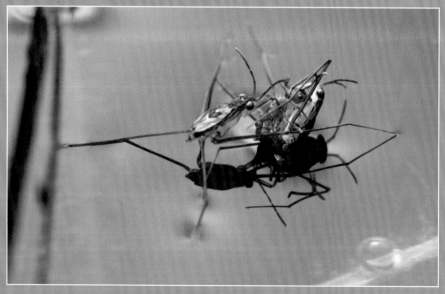

Pond skaters and ant

Reed bunting with dragonfly

Larval dragonfly eating a larval amphibian

Threads

Effective monitoring of a very small population of cryptically camouflaged, semi-aquatic spiders poses many problems. Wading round the flooded pool margins, breaking through encircling and often head-high sedge and reed, provides an extremely restricted view and endangers both spiders and nurseries. To count sufficient spiders to give a reliable measure of their numbers, I rapidly discovered that I needed to be, not on the bank looking in, but down in the pools looking outwards. Searching from the water, close to the meniscus, I could peer into the dark wet tangle of emerging stems at the pool edge to find the poised hunters, the contorted moulters or the slightest rippling message of the amorous. I could look up at the sunbathers stretched full length in the channelled sedge leaves or see below water their aquatic escapes - a tell-tale milky cloud excreted to veil a rapid dive from danger, or the glint of silvery plastron in the filtered sun. Even the sound of a female with her egg sac, dropping from her perch into the safety of water, became familiar. Repeated in a standardised way, this monitoring method gave me not only much higher counts, but also a rare and privileged view of life at this unique interface.

Living at the meniscus presents complex physical, biological and evolutionary challenges. For me, working there, counting spiders, the challenges were more mundane. How to enter the water without sending my subjects scurrying to do the same. How to move forward through waist-deep black ooze, bound impenetrably with a tangle of stoneworts. How to leave again

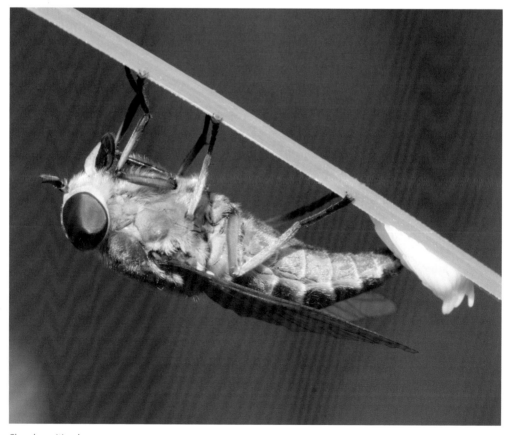

Cleg depositing her eggs

with a shred of dignity, not a crawling, clawing and dripping recapitulation of early terrestrial life emerging from the swamp.

Even without a human audience, I am an object of curiosity or alarm. Reed buntings gather, flicking white tail feathers, bright eyed and complaining on the brown sedge flowers high above me. *Aeshna* dragonflies whirr, iridescent, repeatedly, in my face before accepting my bent back as a perch for their predatory forays and amorous encounters. Alarmed and alarming, the raucous horror of a water rail disturbed from her clutch of delicately pink-spotted eggs in her water's edge nest.

For the blood-feeders of the fen, I am a gift. During the droughts of the 1990s the Fen's production of biting insects all-but-dried up. It was several years before I encountered my first horsefly, beguiled at first by its picture-wings and profoundly green eyes, then hurt by the betrayal of inflicted pain. Even now I admire the eyes and allow a pang of guilt before I swat. The alighting clegs, grizzled grey, wings folded over back on landing, win no such delay. Mosquitos, dark pursing clouds on my walk to work through the encircling woods, now emerge from the water below me, translucent, fragile and at first harmless. For the most part I am protected from the biting females by chest waders and the long sleeves and gloves needed to guard against the thousand cuts of saw sedge. The war paint of mud hand prints, though, slapped on my face in self-defence, records their success in finding uncovered skin.

Caddis case of shell fen molluscs

The greatest reward in this place is to be ignored. These moments leave unforgettable images, in sight, sound and smell, of life continuing at the interface around me. An expressionless frog moves backwards, slowly and smoothly away from the pool margin before vanishing into a mossy hole, its leg gripped in the unforgiving jaws of a grass snake. A lizard, unexpected in the wet fen, gazes up at me before diving deep into the water. Water voles go noisily about their business in neighbouring pools – splashes of disturbed water and crisp crunching of new reed shoots. Below water, occasional flashes of silver mark the departure of water shrews, their presence more usually signalled only by their middens of shell-encrusted caddis cases abandoned in the sedge tussocks. At the water margin a drinker moth caterpillar, victim of a parasitoid wasp, writhes as ranks of tiny white cotton-wool-encased pupae emerge, grizzly, like suckling piglets along its underside. Minute ink cap toadstools, delicate and deliquescing, burst from the wet bases of dead reed stems while even stranger fungi, white and horned, encase the tiny corpses of Clubionid spiders that frequent the dense wet vegetation of the water margins.

Chapter 3
A LIFE IN STAGES

Spiderlings in dew-filled nursery

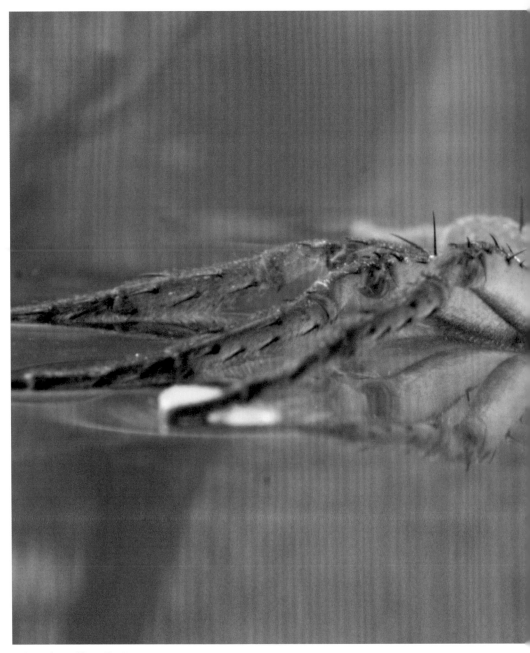

Male spider and his reflection

Growing

From first emergence from a translucent egg, soft and spherical, to the moment of reaching adulthood, the spider's life is a journey in stages. Its form is defined, not by an internal skeleton, but by a tough external cuticle (the exoskeleton). The body is divided into two parts; the front, from which the legs arise, is a fused head and thorax (the cephalothorax) enclosed in two hard, chitinous plates; the carapace above and sternum below. A narrow and delicate stem

(the pedicel) links the cephalothorax to rear of the body (the abdomen) which is enclosed in a softer cuticle.

The confines of this external cuticle set limits to growth. Once reached, the spider must become the ultimate escapologist and extricate itself from its confines before growth can be resumed – a process known as moulting or ecdysis. At each of these moults, the tight old cuticle is shed to reveal a new one waiting soft and wrinkled beneath, giving a brief window of oppor-

12 THE MOULT
Technique: Digital image
11.7" x 16.5" : 3000mm x 420mm

The fen raft spider's exoskeleton is a rigid structure that must be shed many times throughout its lifetime to allow growth. The moult is a complicated procedure with eight legs and a large body to extract from a casing and the method of evacuation is systematic and follows the same pattern for each moult.

The spider will hang suspended from a leaf or stem by a thick silken thread and release itself in a beautifully choreographed series of movements. The top of the cephalothorax, called the carapace, disengages allowing the body and legs to ease themselves slowly free. The body and legs will gradually expand its new skin and fine structures such as jaws and mouthparts are exercised as the cuticle hardens whilst the spider continues to hang supported by the silken thread.

For the spider, moulting is a period of high vulnerability as it hangs helplessly and sheds its body armour. It is at great risk if a limb fails to completely extricate itself from its old skin as it will harden into a disfigured position and impede movement or worse still prevent release from the spent skin. By contrast spiders can often be seen with a limb completely missing yet are able to hunt or reproduce successfully. Missing legs can regenerate in future moults but not always to full size.

These shed skins were collected from spiders and their spiderlings reared in captivity and nurtured in test tubes. Studying them under the microscope it is stunning to see the perfect replica of the original animal including jaws, mouthparts, legs and joints, hairs, spines and casings that cover the eyes.

As the largest of these skins was freed following a moult, the opportunity was taken while it was still malleable, to spread it flat, so that it hardened into this strongly defined shape, instead of the usual crumpled ball of legs and body casing.

Collected skins

Spider shedding its skin

tunity for the compressed body to expand and stretch the carapace and legs. As the carapace hardens, the cuticle of the abdomen of necessity remains more elastic, accommodating the periodic arrival of ingested food and continuing to lengthen between moults.

The spiderling's first moult takes place while it is still packed with hundreds of its siblings in the silk sac in which it hatches from the egg. It is almost impossible to know how many times the spiders moult during their life in the wild; the simple fact of periodic loss of the entire cuticle makes it impossible to mark and track individuals throughout their lives. In captivity at ambient temperatures, spiders from the Fen moulted between eight and twelve times between emergence from the egg sac and adulthood, even in similar rearing conditions.

The process of moulting is of necessity rapid – the spider is extremely vulnerable to predators during this period. In the days before moulting it stops feeding; the abdomen is already round, the cuticle stretched tight. In the hours before the moult begins the spider becomes sluggish, eventually spinning a silk pad on a sheltering leaf from which it hangs, torpid, by its feet and spinnerets, as old cuticle begins to separate from new. The first split is between the carapace and sternum, the former hinging upwards from the pedicle. The abdomen follows before the more challenging task of extricating the palps and eight legs, drawn like long fingers from a tight glove. The newly emerged spider hangs momentarily; its legs at first in a straight bunch and then repeatedly flexed to make the joints supple. When small, only a few brief minutes of exertion separate the appearance of the first split in the old chitin and the emergence of the new body, fresh, fully expanded and ready to face the world. In later moults, escape takes proportionately longer, extending to over thirty minutes as adulthood approaches. At first, even after the freshly emerged spider resumes its normal activities, the new cuticle is soft and the legs have a characteristic green translucence.

The threat of predation is not the only risk during the helplessness of moulting. The summer margins of the fen ponds are strewn with the tangled wreckage of invertebrate bodies that have failed this most demanding of tests. Wind, rain and sun, or errors within - physiological or genetic, unseen - all take their toll. Round eyed dragonflies stare upwards, crumple winged. Caught by a sudden gust of wind on their metamorphic journey, the water that nursed them as nymphs becomes an untimely grave. And mosquitos, delicate, white, and unable to shake off their clinging nymphal skin, will never take flight to join the whining swarms that torment the grazing cattle. For the spiders too, errors in extrication from the old cuticle are costly. Failure to extract completely any part of the body is fatal but death comes slowly, the animal hanging in futile struggle until its energy is exhausted. Even successful emergence, completed too slowly, leaves limbs as useless and curled impediments. Deformed and injured limbs are unlikely to be extracted successfully during subsequent moults; the spiders have been observed biting them off before moult begins.

Limb loss, usually resulting from errors during moult or a too-close encounter with a predator, is rarely fatal and, remarkably, lost limbs can be regained. A new limb begins to form in the remaining basal segment of the old, ready to be unveiled - a perfect miniature - at the next moult. Growing with each successive moult, it is eventually indistinguishable from the other legs. For adults though, regeneration of limbs is no longer an option and they are frequently found with less than eight legs.

The penultimate moult of the spider's life reveals the changes of adolescence. For females at this sub-adult stage, the rudiments of the developing genital opening are visible beneath the abdomen. For males the change is more overt. The palps, until then identical to those of the

Male palps

Final moult

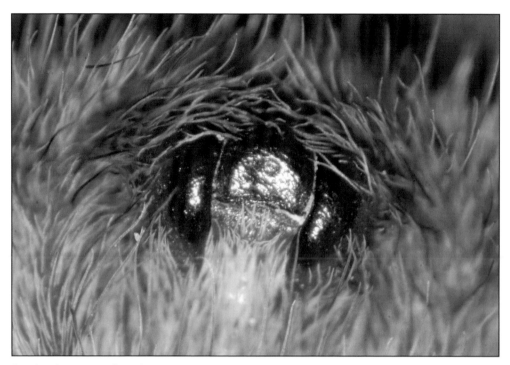

Female epigyne – sexual opening

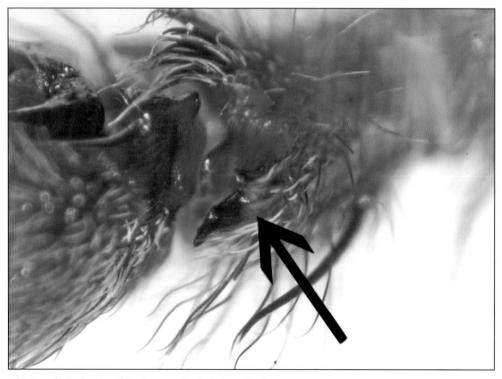

Tibial apophysis: the identifying feature of male *Dolomedes plantarius*

Male palp – expanded sexual apparatus

female, emerge clubbed at the tip – a particular challenge for extraction from the straight and narrow confines of the old palpal casing during moult. These clubs, threatening like a pair of boxing gloves in front view, are rudimentary sex organs.

Adult Life

In both sexes, the final moult, usually undertaken in the spring of their final year, reveals the adult form. The detailed structure of the female's epigyne - a hardened plate now covering the entrance to her fully-formed reproductive tract - is unique to every species of spider and the single most important feature used in identification. In mature males, the body shape is changed. The legs are proportionately and conspicuously longer than before, the abdomen narrower and carapace wider. The bulbous tip of the palp is now revealed as a copulatory organ, its structure unique to the species and, like the female epigyne, providing the primary means of identifying male spiders (thus, most spiders cannot be identified until they are adult). Sperm transfer in spiders is highly unusual with, effectively, two separate acts of seminal discharge. The palpal organ does not produce sperm but serves to store it, and later to ejaculate it, into the female's epigyne during copulation. Much of its complex structure is internal but external protrusions – aphophyses – on the palpal segments help to guide and facilitate the process of mating. It is the structure of one of these apophyses (the tibial apophysis) that provides the most reliable feature for distinguishing the males of *Dolomedes plantarius* from those of *Dolomedes fimbriatus*.

13 COURTSHIP
Technique: Digital print
11.7" x 16.5" : 300mm x 420mm

The fen raft spider's growth to maturity usually takes place over a period of two years on Redgrave and Lopham Fen and the spider is fully mature by its third summer. At this point the spider becomes sexually mature and the fecund female releases pheromones to lure suitors. Males follow a thread of silk she leaves in her wake, a dragline, or airborne vapours picked up by sensory receptors on their palps and legs.

The female's sexual opening is called the epigyne and is found on the underside of her abdomen. The male uses his secondary sexual organs - the palps - which have an enlarged tip, to transfer seminal fluid to the female. He does this by constructing a triangular web into which he deposits the fluid. He then reaches down beneath him with his palps and takes up these cells using an impressive convolution of pipes and ducts.

The male communicates his presence and interest to the female by gently tapping on the water surface then bobbing his body with increasing vigour sending a pulse through the water. He approaches slowly with jerky movements then, in a brave moment, will move to brush against her - risking an aggressive, potentially fatal response. However if this is favourably received she too will join with him in agitated body movements, stimulating him into frenzied bobbing of the abdomen and impressive high flicks of his front legs, as he dances invitingly for her favours.

The female's compliance is assured, and indeed with sixteen legs between them mating is only achievable, when she raises hers to tuck them tight against her body. The male approaches and must roll her onto her back to access her sexual opening and achieve copulation. This is a difficult procedure with even the most compliant female - legs hooked but pushing against a liquid bed. Once this has been achieved, the male transfers seminal fluid from each palp in turn into the female's epigyne.

Close observation of pairs of spiders during courtship and mating provides a wealth of information about their ritual dance and essential anatomy. This was fascinating and gave a sense of wonder at the drive and energy each individual expends on reproduction as well as the hugely complicated and unlikely apparatus nature has provided for the purpose.

Understanding the process did not directly lead to an image that could successfully represent the impression of confrontation and confused animation that is displayed immediately before they mate: tense, dynamic and full of action.

This blurred, low - resolution image was chosen and then re-photographed direct from a screen to lose more definition, before being manipulated further to obtain a more even graininess and balanced contrast. The image was then made into a photographic etching using a photopolymer process and printed to a fine art specification which allows more depth and lustre than a photographic print.

High leg flicks

following a dragline	agitated body movements	high leg flicks	female	touching legs	female drawing legs in	rolling the female over

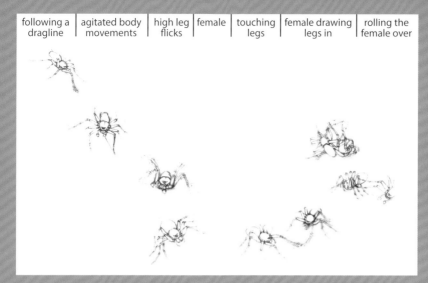

Courtship dance

Following the dragline

Copulation

The sperm are produced in testes coiled within the spider's abdomen, and secreted through two inconspicuous ducts. Soon after the final moult the male spins a tiny, suspended triangle of silk - a sperm web - onto which he discharges a drop of semen. Dipping his palps alternately into the drop, he charges a reservoir inside the palpal organ, priming it for his new life - a dedicated search for mates.

A pulse of ripples, small but distinct amongst emerging sedge stems, creases the flat blue water of a May morning; a male spider has begun his courtship. The object of his attention is hidden deep amongst the emergent sedges, formidable, her heavy abdomen encasing a waiting clutch of eggs. Lured by her pheromones, at first subliminal in the breeze and then tangible, impregnated in her delicate silk drag lines trailed across the pool surface, he tracks her with chemosensory hairs on his palps; her presence is literally palpable. Perhaps because the trails are many and confusing, or the pheromonal signals insufficiently precise, final determination of the female's precise location seems difficult for the courting male. Not until she responds to his patterns and pulses of ripples, sent first as tentative enquiries and then more assured and confident, does he seem sure of his goal. The courtship is often protracted, ritualistic, constrained perhaps by the potential high cost of misjudgement. Although amongst almost a hundred courtships observed in captivity, only one ended in cannibalism, the possibility of this ultimate price appears to select strongly for cautiousness.

Gently tapping or sequentially quivering his front two pairs of legs in delicate arcs across the water surface, moving his abdomen from side to side, and slowly bobbing his whole body, the male sends a sensory barrage of signals to the waiting female, broken only by pauses for essential grooming of his legs and palps. If receptive, the female signals assent either by passivity or by replying with slow body bobbing at a late stage in the courtship; vigorous bobbing at an early stage, violent upward flicking of her front legs, or a measured, slow-motion walk away from him, constitute a firm rebuff. Eventually in close proximity to the female, the increasingly animated male adds visual cues to his repertoire. His advance is signalled by a series of rapid, powerful upward flicks of his front legs. Then, finally making physical contact with her, his delicately dancing legs brush over her body. She slowly retracts her own legs, close to her flanks, in reply. With the daunting obstacle of eight long and powerful legs removed, the male interlocks his legs with hers and, still with great physical effort, pulls her over in the water to access her epigyne on the underside of her abdomen. Hitting it with his palps, the complex, tightly packaged structure of his palpal organ locks in and explodes, the droplet of seminal fluid often clearly visible. After a split second's pause the male leaps back, once again flicking his front legs upward defensively as he frantically grooms to repackage the palpal organ. Sometimes he manages to discharge both palps at the first approach but often two approaches are needed – a double risk.

The males court and attempt to mate with multiple females, often alternating their attentions between several at once in densely populated areas, sometimes hunting over long distances for isolated loners. Several males may stake-out a single female with little evidence of overt aggression between them, their precedence apparently settled by more subtle signals. The lure of females is so great that the males lack discrimination, gathering around those already carrying egg sacs, pursuing the still sub-adult and even tracking down the recently dead. As the females swell with eggs the males shrink, their ever narrowing abdomen testimony to the demands of their single-minded pursuit, fuelled only by occasional, half consumed meals. By early July most have succumbed to starvation and exhaustion.

Gravid unbanded female

The egg sac

The female's abdomen grows heavy with her burden of eggs, her skin eventually so stretched that it shines. Broods of between five and six hundred eggs are common though occasional broods exceed seven hundred. In the twenty-four hours before she lays her eggs, the female's behaviour changes. She becomes sluggish, as if approaching a moult. Vulnerable now to predators, she hides deep in the dark and cavernous base of a sedge tussock, hanging from the black roof, torpid. And then begins one of the least witnessed and most remarkable events of her life. Becoming active again she raises her abdomen to the roof above her and spins a silk pad. Then, pivoting in slow circles she repeatedly raises her abdomen – attaching the silk that now streams upwards from her spinnerets like beams of light – and then lowers it, stretching the next skein of emerging silk. Through this revolving cycle of stretching and attachment of silk, she spins an inverted cup, a cradle of the whitest silk, dense and crinkled. Pausing briefly, her abdomen contracts and pushes upwards, filling the cup with eggs, imparting a yellow yolky glow. Then, pivoting once again, her abdomen now deflated and wrinkled with new emptiness, she once more streams silk upwards to encase the eggs, seal the cup. And when this ball of cotton-wool silk and eggs is complete she cuts it free of its anchoring threads and turns it with her legs, adding bands around it, no longer of wrinkled silk, but of tight threads, strong and binding. When complete, her chelicerae, once venomous in attack and defence, become cradling hands clutching her precious creation. From this moment the female her egg sac are inseparable. Her investment is great and her need to capitalise on it absolute.

By the time she is seen on the water again, her sac is as muted and brown as her surroundings, the bright white silk oxidised by contact with water. Up to one and a half centimetres in diameter, the sac is a cumbersome burden requiring the mother to walk on tiptoe and making

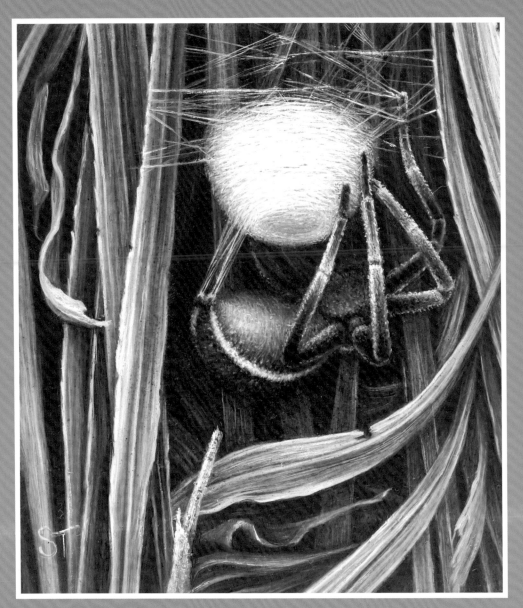

14 Spinning the egg sac
Technique: Oil painting
4.5" x 4": 115mm x 100mm

The abdomen of the female fen raft spider swells with the development of eggs as the time nears for them to be laid into a special silken container that she must build. Climbing free of the water surface to a safer, secluded position amongst the vegetation she anchors supports for the egg sac construction on strong stems and builds a scaffolding from silk whilst hanging upside down.

Liquid silk is produced from ducts in the spider's abdomen. It passes through small, muscular appendages called spinnerets at the end of her abdomen. Silk is a protein and the thread has unique qualities: it is stronger strand for strand than steel yet elastic and flexible. It can be varied in thickness and texture according to its particular function and has excellent adhesive qualities.

The spider gently moves her abdomen up and down, releasing silk, weaving a spherical silken cup whilst suspended under the rigged scaffolding. Pressing her body tight against the sac she expels the eggs into it, and her previously swollen abdomen visibly shrinks. The opening of the cup is sealed over and the sac cut free, before a further dense layer of silk completes the task of providing protective layers for the fragile cargo of eggs.

Research for this image involved the powerful and privileged experience of watching this tiny creature instinctively construct a beautiful cocoon using materials sourced from her own reserves and specialist body parts. The strands of silk that were pulled through her spinnerets during each successive anchoring caught the light from the camera's flash and gave solidity to the shaft of silk in the photograph.

In the painting we enter the privacy of the spider's miniature world deep within the clump of vegetation as she seeks protection from any possible threat to her fragile creation.

The last egg

Female with swollen abdomen

A shaft of silk

15 FEMALE WITH EGG SAC
Technique: Wood engraving
3" x 4": 75mm x 100mm

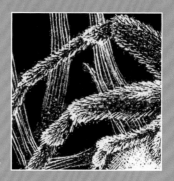

Female with egg sac detail

The female fen raft spider carries the egg sac clutched securely against her body for around three weeks, using her chelicerae and palps. The eggs develop and hatch into tiny spiderlings during this time, and undergo their first moult in these safe but confined conditions. She must stay near to water throughout this period: moisture is essential for the development of the spiderlings and the material of the sac, and she frequently dips it into the pools.

The female spider might not feed during this period of time and is much more guarded about being exposed with a restricted mobility from carrying her precious sac. She will quickly disappear from view at the approach of any perceived threat.

If there has been no successful mate she will produce a virginal sac with infertile eggs. Mature males rarely survive beyond mid to late July, when their job is complete. Meanwhile, the female stores her bank of sperm in order to produce a second fertile egg sac in late summer.

Wood engraving is a relief printing process that uses hard end grain pieces of wood for the printmaking plate. Special fine, sharp tools are used to cut away areas of the wood to create a drawing.

The surface is inked and the print is made by placing fine art paper onto the plate and rubbing the back of it by hand or putting both through a relief press.

It is the most precise of all relief print methods as wood for the plate is taken from slow growing species such as box or lemon and the surface is prepared to the smoothness of polished marble. The minutest of marks register in the print, giving great opportunity for fine detail and texture,

This intimate wood engraving aims to convey the determination and strength of the female as all of her reserves are focussed on the safe and successful nurturing of her young. Her whole body embraces the sac, and she moves at one with it during the weeks that it is carried.

Female with egg sac near the water margin

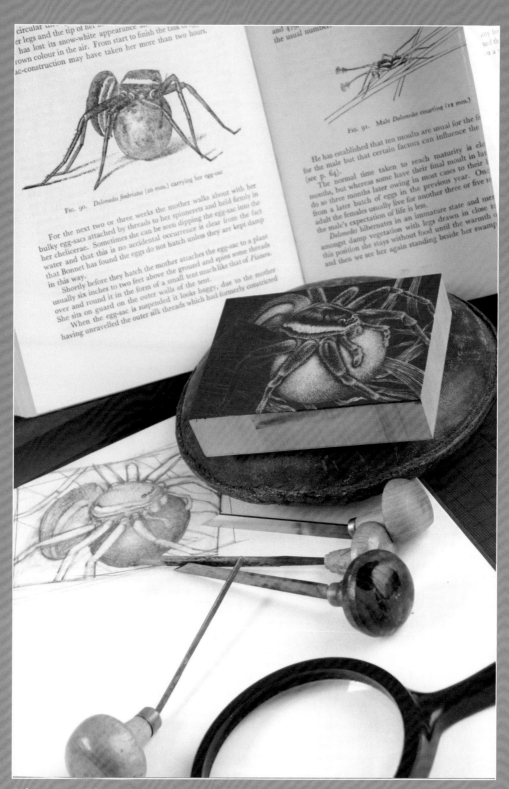

Reference material for print

Female immersing egg sac

her vulnerable. Carrying it for the next three to four weeks, she is loath to release her hold on it, often fasting for the duration. Regular grooming to maintain her all-essential waterproofing is the only task for which she will briefly relinquish her grip but even then she cradles the sac firmly with one leg, seizing it up again in her chelicerae and at the slightest hint of danger. In warm weather her days are marked by the need to keep the sac moist. Spending her time near the water's edge, she descends every few hours to sit on the surface, completely immersing the sac below her body to maintain vital humidity around, first the eggs, and later the hatched young contained within it. As the time approaches for the young to emerge, the mother's routine changes. She spends more of her time higher up in the tangled saw-edged sedge leaves, the spot chosen for her nursery, though still making the precarious pilgrimage down to the water to immerse the sac.

The silk sac is both cradle - softly silk-padded - and fortress - impenetrable and almost impossible to pull apart. But in the week before the spiderlings emerge, it becomes softer, no longer a firm sphere but a sagging sack. It is not known whether the spiderlings play a role in facilitating their own escape from the sac but mother's role appears to be instrumental. In the safety of darkness, she turns the sac slowly in her feet, punching through the silk with her powerful chelicerae. By morning the exterior of the sac appears peppered with bullet holes, escape routes for the tiny spiderlings inside. Over the next twenty-four hours a few tentative escapees emerge, often sitting on their mother like tiny parasites. But the mass exodus, the bursting free of the tightly packed young, is a private affair, usually reserved for the quiet grey hours before sunrise, safe from the prying eyes of diurnal predators.

Climbing female with egg sac

Female poised above the water surface with egg sac

Emerging spiderling

The nursery

Slanting shafts of early morning sun light up a new and glowing beacon in the sedge. A dense and dew-filled sphere of silk hangs high above the dark water, firmly anchored to the leaves by long threads. This has been family work, the mother providing the framework and the spiderlings themselves, with every exploratory venture, contributing their silk drag lines to the dense and protective matrix of the nursery. Clearly visible within the web as the dew melts away, the empty silk egg sac, torn and distorted where the spiderlings disgorged into their new home. And dragged along in the rush, trailing from the gaping exit, the tiny translucent skins left behind at their first, hidden moult. Within the nursery the spiderlings are packed again, spherical, as if the confining sac were still present around them. Watched carefully, this spiderling sphere flattens to a concave lens and pans slowly round, tracking the sun across the summer sky and harvesting its warmth.

As dusk gathers, the ball of spiderlings expands and then dissolves in the grey light. The tiny spiders move out through their web, thickening and extending it, some intercepting for their first meals tiny feather-winged thunderflies caught in the mesh. Until this time they have been nourished only by the diminishing reserves of the yolk within their abdomens. Their frequent encounters with their siblings – a ritualised and polite waving of front legs – hold no danger. The mother guards her brood, waiting, hidden, stretched below a stem, in contact with the threads that send her telegraphed warnings of danger. In the gathering beat of rain she moves to

16 NURSERY WEB
Technique: Collograph
24″ x 6″ : 610mm x 150mm

Nursery web detail

As the fen raft spiderlings develop in the egg sac and the yolk reserves are depleted, it is time for the female to create a nursery web, dry, up above the watery level of the pools, for the next stage of their lives.

She builds a dense cluster of fine silk threads amongst the rigid stems of sedge and agitates and turns the sac, puncturing the surface with her jaws. Eventually this enables her young to climb out onto the delicate mesh of the web where they will clump together, exposed for the first time to wind, sun and rain. As confidence builds they discover the outer limits of the web adding their own threads, returning to regroup again, forming a ball of tiny fidgeting forms. Meanwhile the empty nest of the egg sac remains suspended in the web.

Here they stay, closely guarded from predators by the fiercely protective female, who shields them with the broad shape of her body. After six or seven days the spiderlings are mature enough to leave the sanctuary of the web to begin an independent life of their own.

This plate again employs some of the leaves used for the Sedge image (Image 5) but combines them with cut shapes of card mimicking the array of broken and dead leaves found at the water level. Overlapping threads create the nursery web cluster suspended above the water surface as it laces across rigid supports. The powerful silhouette of the female spider, overviewing her brood high on the sedge, is visibly exposed; she is prepared to make the ultimate sacrifice for her offspring.

Emerging spiderling

A nursery web on the Fen

A cluster of spiderlings

Torn sac

place her body between the cluster of spiderlings and the force of the drops. The sharper shock of a predatory beak triggers a more violent response. Rushing towards the aggressor, the mother is fearless, fearsome, and attacking while behind her the spiderling ball explodes and vanishes, the tiny spiders spilling down the rigging of the nursery to the relative safety of anchoring leaves. In seconds, a substantial meal for the predator has fragmented into mere crumbs.

The pull of motherhood is increasingly countered by the need for food after so many weeks of near-starvation. Using sperm stored from her early summer mating, the mother can produce two broods during the East Anglian summer and the need to replenish her reserves to prepare for the second must be weighed against the benefits of vigilance for the first. In quiet moments she slips down to the water's edge, unburdened, poised again for the catch. As their first week nears the end, fewer of the spiderlings return to the safety of their sibling cluster after their night-time forays. Their protected nursery days are over and they vanish into the damp waterside vegetation to begin an independent life, fraught with danger. In captivity, broods left together beyond their normal dispersal time turn increasingly to cannibalism, the niceties of ritual greeting supplanted by the need to survive.

The empty nursery clings on, sometimes remaining conspicuous and largely intact for two or more weeks, although, within a few days of the spiderlings leaving, the dense internal structure

Close neighbours

17 MOONLIT NURSERY
Technique: Wood engraving
5" x 6": 126mm x 152mm

The nursery web provides a protective environment for the first week of the spiderling's life outside of the egg sac. It has been over a month since the eggs were first cocooned into the egg sac and now the spiderling is ready to venture out into a wider world and test its ability, relying on the uncertain provision of life on the Fen.

An instinct for independence spurs it to send out a thread of silk that lifts with air movement and attaches to nearby vegetation. This becomes one of many highways for the great exodus and now the nursery empties of hundreds spiderlings. They disperse into adjacent vegetation to start a life foraging on their own.

The spiderlings face many trials as they journey to adulthood and a successful conclusion to their life cycle. Variations of food supply, extreme weather conditions or deteriorating habitat, and an abundance of creatures who view them as a tasty morsel, are all possible dangers that lie ahead.

The ability of a wood engraving to register fine marks while retaining a densely black background allowed this piece to convey the half-light of a nocturnal scene, with softly expressed shapes caught in pale moonlight. Here spiderlings move with the cloak of night offering some protection from predation. This is starkly contrasted as they move towards the day and their tiny, dark silhouettes are revealed against the clear, bright sky.

Final days in the web

Female on guard

Spiderlings dispersing

Old female in empty nursery web

starts to break down and no longer gathers the morning dew. It is easy to gain an impression that nurseries endure far longer than this because other females with egg sacs may find safe refuge, and later build their own nurseries, within the protective curtilage of recently abandoned, or even still-occupied webs. Second brood nurseries are also built very close to the site of the first, making accurate estimation of nursery numbers a complex challenge.

For the mother, the demands of rearing her brood take their toll. Relatively few seem to be successful in producing second broods on the Fen, although in the ditches of the Pevensey Levels there is evidence that some go on to produce a third. Conditions on the Fen in summer are less dependable than on the grazing marshes. When rain fails the pools dry out rapidly leaving the emergent sedges on which the spiders rely for nursery construction stranded above the receding water's edge. Predators gather around the diminishing resource of flooded pools to quench their thirst and exploit the ever increasing concentration of aquatic prey. The matriarchs that survive these challenges to produce a second brood often bear the scars of their exertions – the plump abdomen now collapsed and angular, limbs lost and a once immaculate body now water stained.

For those successful in rearing a second brood there is no urge to return to the water and replenish. There seems to be no option for surviving a further winter, just a slow fading with the shortening days. Many of these females simply remain with their nurseries until they die, some dying even before their young disperse, others sitting out many more autumn days.

Dipwell in winter

Immature *Dolomedes* overwintering in dipwell cap - *Photo Arthur Rivett*

Life-span

For the young spiders, the frosts of winter bring torpidity and the challenge of hibernation. Little is known of their hiding places although air spaces in hollow stems and amongst the dense bases of sedge tussocks are likely to be their main retreats. On the Fen, wintering spiders have been found in the hollow metal tubes of dipwells, used for measuring water levels in the peat sediments. Like a large-scale model of a hollow stem, perforated by boring insects, the spiders enter through small holes designed to allow the passage of water, and then ascend to the dry shelter under the capped top of the tube.

Of those that survive this first winter, most continue to grow through the following summer, usually undergoing their penultimate moult by their second autumn. These sub-adult spiders must once again find safe retreats in which to hibernate, emerging again as the first warmth of spring penetrates their hiding places.

After their second winter in hiding, the majority of the Fen's raft spiders mature to complete their life cycle. But like many invertebrates, the more they are studied the more it becomes clear that their growth rates and life spans are very variable. Although there is no evidence to date that fen raft spiders ever extend their adult life beyond a single season, there is increasing evidence of plasticity in the number of years taken to reach maturity: some breed after only one year while others may survive to mature in a third (Chapter 4). This potential for flexibility in their life cycle must have contributed to their tenacity, giving them more capacity to sustain populations through periods of drought and breeding failure.

Threads

The life history of very rare and protected animals is always intrinsically difficult to study and all the more so that of invertebrates which so unobligingly moult away any markings applied to identify them. Much of our detailed, though still very incomplete, understanding of the fen raft spider's life history is a by product of keeping them in captivity for the research and rearing programmes needed to inform and deliver their conservation (Chapter 4). After the temporary

West from the fen

luxury of being able to achieve this through well-funded studentships and laboratories (Chapter 1) came the economic necessity of kitchen science. In the undignified setting of plastic trays adorned with sterile plastic pondweeds, my kitchen witnessed the unabashed intimacies, courtship and copulation of many pairs of spiders. Their every move and intimacy was carefully recorded although my attempts to avoid egging-on, or providing anthropomorphic advice, were largely a failure.

On the Fen, in the water, gathering data on courtship is a vastly more time-consuming challenge. It is encountered by chance, acted out amongst the privacy of dense emergent sedges, and often so protracted that the threat of being irretrievably embedded in mud eventually overwhelms scientific curiosity. But even relatively scant observations obtained in the wild are vital in helping to interpret the plentiful information gathered in an artificial setting. Used together, these sources of information start to build a more complete picture of the spiders' life. Only in captivity have I been able to quantify the elements of courtship behaviour but only in the wild could I be certain that simultaneous courtship of multiple females is part of the males' normal behaviour. Only in the wild have I glimpsed spiders mating on vegetation above the water surface – just once and almost certainly exploiting the delicate defencelessness of a female newly emerged from her sub-adult skin, a behaviour well-known and documented in some

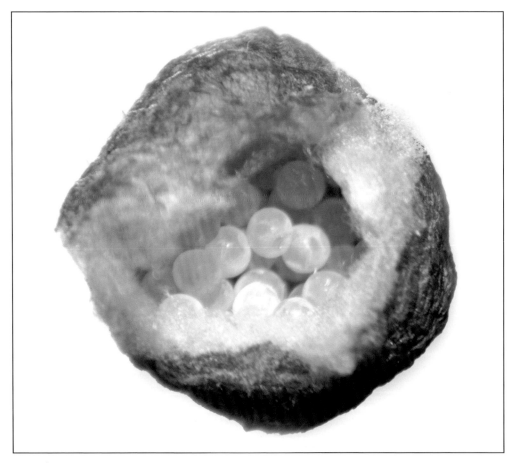

Virginal egg sac

other spider species. Length of life and numbers of moults can be precisely recorded in captivity but lack relevance without some parallel measurements from the wild, however difficult to obtain. The factors influencing them though – temperature and food supply for example - can only be manipulated and quantified within the rigorous setting of experiments carried out in captivity.

I had been working on the Fen for seventeen years before I encountered one of the most amazing of this species' behaviours - the spinning of silk sacs in which to carry their eggs and newly-hatched young. Late one August afternoon, close to the water and peering deep into a hole at the base of an island of sedge, I saw a brilliant white object, incongruous in the muddy darkness. Puzzled at first, I began to see, beneath the white, in the gloom, a familiar pair of raft spider stripes. I was looking at the inverted rear of an abdomen. After a brief pause, as if suspecting discovery, the spider began to pivot and silk streamed in amazing volume from her spinnerets, adding to the rim of an inverted cup of white silk. Eventually stopping again, she faced me. Her eight eyes, modestly protected by long curving lashes, stared at me and I feared my presence had ended the spinning, broken the spell. When to my relief she resumed her work, it was clear that the pause was not of my making but marked the laying of her eggs, now being bound securely into the cup. Her round abdomen was suddenly wrinkled and deflated

Full moon in June

and from the heart of the silk came a yellow glow. I watched until she completed the sac and carried it away from me, deeper into the darkness.

Since then I have witnessed the spinning of sacs many times – late at night, in my kitchen, but never again in the field. Even in the mundane setting of a plastic aquarium in a kitchen, it never ceases to be a privilege to be party to it nor detracts from the indelible memory of that first encounter in the dark sedge on the Fen.

The kitchen has also borne witness to the female spider's devoted privations while carrying their precious sacs. The maternal instinct is strong though not discerning. After their first broods, females occasionally produce sacs with infertile eggs. These are carried long after they are due to hatch, or put down only to be replaced by a piece of torn cotton wool bedding, which is carried with equal tenacity. Once, when a female nearing the end of her life dropped her egg sac, she held in her chelicerae instead the top of a bamboo cane on which she had been perched. In an eventually successful attempt to reunite her with her sac, she relinquished the cane only in exchange for the tip of my finger. The extraction of her chelicerae without causing them damage was a delicate task requiring fine needles and dexterous assistance!

Observations of the nocturnal emergence of spiderlings from their now sagging silk sac have proved more difficult to capture. Although egg sacs seem to be constructed by preference in darkness, the urgency of egg laying means that once begun, the task will be completed even in the presence of some artificial light. Spiderling emergence is more strictly nocturnal. Only once have I watched a mother turning her sac with her legs and palps while punching exits through the silk with her chelicerae. Only once have I witnessed the later stage of nursery construction and mass emergence of the spiderlings. Hurrying to-and-fro across the aquarium, the mother anchored silk strands to the sides as she constructed the framework of the nursery. Seemingly no longer content with small holes as exits for her large brood, she repeatedly interrupted her spinning to rip open the now silk-anchored sac with her chelicerae and open the flood gates.

Chapter 4
ON THE MOVE

Winter fen

Bee orchid on the Fen margin

A Plan of Action

Throughout the bleak years of Redgrave & Lopham Fen's desiccation and decline, targeted conservation measures for the fen raft spiders almost certainly enabled them to survive until closure of the artesian bore-hole in 1999 (Chapter 1). The failure of their population to respond to the re-wetting of the Fen at that stage, and the research this engendered, proved a turning point in their conservation history. The stark evidence that genetic diversity was declining on

the Fen, that the spiders were being irreparably impoverished, underlined the high risk involved in a wait-and-see approach to their recovery. For UK population of this species, all eggs appeared to be lodged in just three baskets, at least one of which was wearing very thin.

Backed by this new evidence, Natural England (formerly English Nature), took urgent action. To help secure a long-term future for the fen raft spider, by now a wetland icon, they embarked on an ambitious plan, not only to bolster the fragile population at Redgrave & Lopham Fen, but also to establish new populations. The plan was underpinned by the Biodiversity Action Planning process,[1] the UK Government's response to the Convention on Biological Diversity - a hard-won product of the Rio de Janeiro Earth Summit in 1992. The Action Plan for the spiders set a challenging target - to increase the number of populations from three to twelve by 2020. The target for Redgrave & Lopham Fen - the re-occupation of much of this population's likely former range - was also ambitious.

The chances of the national twelve-population target being met either by the discovery of overlooked populations, or by the natural establishment of new populations, were extremely low. At the time of writing, the rate of discovery since they were first found at Redgrave & Lopham Fen equates to just one population every twenty-nine years. Nevertheless, to ensure as far as possible that our understanding of the spider's rarity was realistic, targeted survey work in potentially suitable wetlands was increased and a campaign begun to raise public awareness of the search. To date, six years on, no new *plantarius* populations have been found although many new records of the very similar *Dolomedes fimbriatus* have resulted from the search. Although residual populations of *plantarius* may yet await discovery, it is increasingly improbable that there are enough of them to change our view of this species' fragility.

The research commissioned to inform practical conservation measures for the spiders showed that the chances of new populations establishing naturally, by dispersal from the old, were also extremely low. The spiders were too unadventurous in their explorations and the nearest suitable habitat too far away. By marking the spiders' hard carapaces with unique patterns of tiny spots of paint and observing their day-to-day movements, PhD student Phil Pearson was able to show that juveniles and adults ranged over distances in the order of only a few metres. The only common exception to this was the compulsive journeying of adult males as they tracked potential mates across the fen pools. Even tiny spiderlings dispersing from their nurseries failed to use opportunities for longer distance dispersal by taking flight and ballooning to new habitat, choosing instead the conservative option of rigging across their silk tightropes to nearby vegetation (Chapter 3). Further indirect evidence for the spiders' stay-at-home lifestyle came from PhD student Marija Vugdelić's finding that the two tiny remaining populations within Redgrave & Lopham Fen had become genetically distinct after only around forty years of separation by a distance of just seven hundred metres. This was clear evidence that there was no significant interchange – or gene flow - between these populations. Staying put clearly paid. Explorers and aviators were so unlikely to find new habitat in numbers sufficient to establish viable populations, that natural selection removed adventurous tendencies from the population.

If these unadventurous spiders were to establish new populations, it was clear that they would need transport. Radical plans were developed to move – or 'translocate' - spiders from the existing populations to new areas of suitable habitat. The planning process was complex with many questions to be considered. Research had shown that the long-isolated spiders of the Pevensey Levels had come to differ from those on Redgrave & Lopham Fen, probably through differential loss of ancestral genetic diversity as their populations contracted. So, which of these populations should supply spiders for the new sites? If they were taken from Redgrave & Lopham Fen, how could this be achieved without risk to this already fragile population? At what stage in their life cycle should the spiders be introduced, and how many of them, over how many years? Suitable areas of habitat had to be defined, located and checked thoroughly for hitherto undiscovered populations. But if the spiders were successful in these

Tufted vetch buds

new areas, could it be at the expense of other rare species, already present?

A translocation plan emerged gradually from the work undertaken to answer these questions. The sites chosen for the first new populations were nature reserves on the river Waveney, downstream from Redgrave & Lopham Fen – an area where the spiders must almost certainly have occurred in the past, before the valley's habitats were fragmented by urban development and agricultural intensification. As well as this geographical proximity to Redgrave

18 DEGRADATION
Technique: Mixed medium
15.75" x 31.5" : 400mm x 800mm

The ever-increasing demands asked of post-war agriculture to raise production levels, coupled with the opportunities that mechanization brought to farming, meant change increasingly swept across the East Anglian landscape from the middle of the 1900s. Hedging and heathland was ploughed and opened up to the larger-scale mass-production of food. Much of the grazing marsh edging the Waveney was drained and given over to crop production. At the same time, water extraction was increased to meet the growing demands of farming and people.

The introduction of intensive farming methods and reliance on the agrochemical industry from this time led to a trail of pollution from livestock slurry, pesticides and fertilizers that ran off into waterways. These practices had a devastating impact on the wetland habitat of the Fen, and *Dolomedes*, so newly discovered in 1956, was increasingly under threat.

This image is a map of the River Waveney, as it meanders snake-like along a twisting channel from its headwaters at Redgrave and Lopham Fen and drains from West to East towards the sea. Washes of colour, overlaid in a toxic mix, represent the land being stained from farm use, filtering through the ground into the waterways that leach their discharge into the Waveney.

Collage sketch

Pig farming

Reflections in stained puddle

Wheat field

Meadow thistle - July

& Lopham Fen, these reserves offered the potential for the new populations to expand over a wide area. Lying at the heart of Suffolk Wildlife Trust's 'Suffolk Broads Living Landscape Area', they were surrounded by an increasingly large and interconnected expanse of restored grazing marshes. If they thrived at their release locations, eventual recolonisation of this extensive area would add value to the translocation and increase the likelihood of developing large and resilient populations.

As on the sites that are home to the three natural populations, the aquatic life of these ditches was amongst the richest in the UK and included a similar suite of national rarities. It seemed likely that here, as there, the spiders would slot into this web of life; as generalist predators they had the potential to strengthen it by adding to its complexity rather than weakening it through their predatory activities.

Although the habitat on these seaward marshes of the Waveney appeared to be ideal for the spiders, it did so by virtue of similarity, not to the nearby sedge beds of Redgrave & Lopham Fen, but to the grazing marsh ditches of the Pevensey Levels on the East Sussex coast. This raised a dilemma over whether the spiders should be introduced from the more local site or the more similar habitat. Because of the genetic differences between these two populations, there were also good arguments for establishing new populations using spiders from both of them; this would provide natural selection with the greatest choice of genetic variation from which to fine-tune the new populations to their habitat and to future threats from our changing climate. But the arguments were more complex still; there was a further reason why mixing the populations could be an advantage. Genetically isolated populations of this sort can sometimes show what is known as hybrid vigour when they are brought together – their offspring thrive better than those of either parent population. Conversely though, offspring of mixed parentage can potentially do less well than their parent populations – a phenomenon known as outbreeding depression.

To help resolve this complex dilemma, an experiment was carried out to test the consequences of mixing spiders from Redgrave & Lopham Fen and the Pevensey Levels. Crosses were made between adults from the two different populations. Using rearing techniques developed at John Innes Centre Insectary in Norwich, the survival and growth rates of their offspring were compared with those from pairings where both parents came from the same population. The results showed that spiderlings with parents from both Redgrave & Lopham Fen and the Pevensey Levels fared neither better nor worse than those with parents from a single population – maximising the genetic diversity of the new populations by using both sources appeared to be the best plan.

Travellers

By 2010, plans were in place for obtaining the spiders needed for the new populations. On the Pevensey Levels, where the ditches are carpeted green with the spiky rosettes of water soldier, the population is so dense that removal of small numbers of spiders is likely to be compensated by improved survival of others; with all the necessary licencing and consents in place for this highly protected species,[2] the spiders could simply be harvested for translocation. But at Redgrave & Lopham Fen harvesting posed too great a risk to the fragile population and obtaining spiders for translocation presented a greater challenge. The captive rearing techniques already developed held an effective though labour intensive answer, with high survival rates delivering vastly more spiderlings than would come through this particularly vulnerable period of their

19 BOTTLENECK
Technique: Linocut
17.75″ x 12″ : 450mm x 305mm

By the 1990s the fen raft spider population was severely threatened by water abstraction and habitat degradation, resulting in a severe decline in numbers.

This prompted concerted conservation action for the spiders through the Species Recovery Programme in 1991. Later in the 1990s work also began to restore the Fen's rare habitats. Tree and shrub encroachment was stripped out and degraded peat was removed before water abstraction was finally ended in 1999.

However despite restoration of their habitat the spiders failed to thrive and research showed that genetic diversity in the population had declined. Severe reductions in population size, resulting in irretrievable loss of genetic diversity, are called bottlenecks. If bottlenecked populations eventually recover in numbers, genetic diversity remains low and this can decrease the ability to adapt to changing environmental challenges.

The shape of a bottle became the dominant theme of this image, aptly describing - visually as well as verbally – the constraint placed on Redgrave and Lopham Fen's spider population. It stands architecturally in the central structure of the composition with pockets of spiders trapped in separate shapes abutting against each other. They are unable to traverse the confining boundary that isolates them within their individual territories and the population diminishes as the shapes descend to the neck of the bottle.

Initially the image was planned using colour to emphasise a loss of vitality – or a lack of genetic diversity - during the inevitable decline into the neck of the bottle. This had worked well in sketches, however the roughly torn edges and natural fibre of this handmade paper reflected the vegetation of the Fen more closely. Although colour would have added a dynamic element to the image, black and white reads as a more sombre definitive statement.

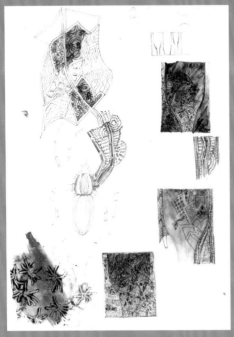

Sketchbook studies I

Sketchbook studies II

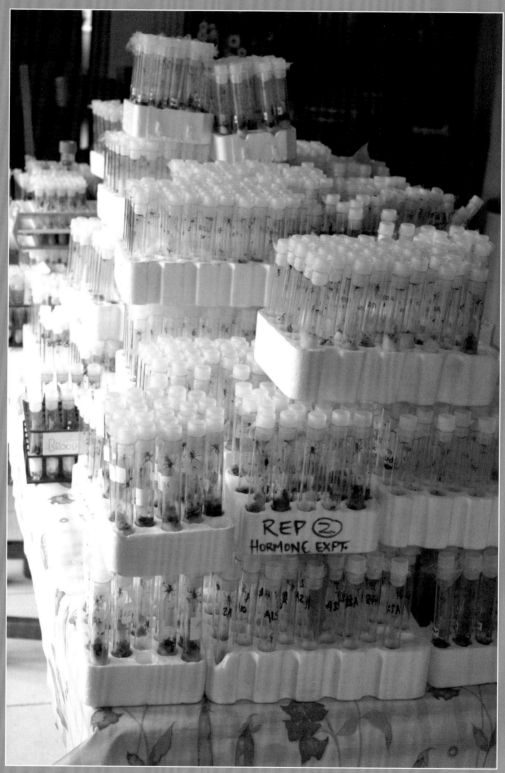

The kitchen table

lives in the wild. Spiderlings from the Fen could be mass-produced for release at the new sites and at the same time the Fen population could be augmented rather than depleted.

Adult female spiders carrying egg sacs were collected from the Fen and kept until they hatched their young and built their nurseries in plastic water bottle 'aquaria'. After a week, when the tiny, still translucent spiderlings would normally abandon the nursery, they were carefully transferred to individual test tubes. In racks and ranks, these became their home

Water soldier carpeting ditch at Calton Marshes

for next three months of their lives. Safe from the risks of sibling cannibalism and predation, and supplied with a diet of tiny, live flies, they grew fast and survived well; in many broods well over ninety percent survived. This production line delivered many thousands of spiderlings from a just small number of females removed from the Fen. By early autumn the Redgrave & Lopham spiderlings, now around four millimetres long and perfect miniature replicas of their parents, were ready for release at the new sites. Rearing them in captivity beyond this

Captive spider with spiderlings

stage, necessitating larger housing, larger food and further research on their requirements during hibernation, was impracticable and the survival benefits were lower.

The release, just like the production, of the test-tube spiderlings from Redgrave & Lopham Fen was labour intensive. The test tubes were strung, well-spaced, along lines of tape distributed amongst tall ditch-edge vegetation; releasing the spiderlings in one spot would

have risked cannibalism and made an easy feast for observant predators. Once the caps were removed from the tubes, the spiderlings were free to venture into their new environment, away from the cotton wool safety, close confines and assured food supply of the past three months.

While the Fen spiderlings acclimatised in new homes, their mothers remained in captivity. Research at the Fen showed that they were much more likely to produce successful second

Test tube babies

Strung out and ready for release

Transportation to new site

broods of spiderlings in captivity than in the wild. To exploit this opportunity to boost the wild population, the females were retained until they produced their new nurseries, and only then released back to the Fen complete with their water-bottled family. The bottles were simply anchored in suitable pool-edge vegetation, opened and left for the spiderlings and mother to venture out.

By contrast, the harvesting of spiders from the large Pevensey Levels population was a much simpler process. Females with egg sacs were collected from ditches supporting very dense concentrations of spiders. But rather than moving them directly to new sites, they too were retained a little longer in captivity; direct release of mothers carrying cumbersome sacs packed with hundreds of unhatched spiderlings would have made a single mouthful for a hungry predator. Instead, as soon as the spiderlings were ready to abandon nursery life, the entire bottled family was transported to the new site for release.

Taking chances: banded spiderling ready to go

Newly released spiderling confronts a Tetragnathid spider

Water bottle aquarium at Carlton Marshes

Spiderlings leaving a bottled nursery

The return of the natives

Broadland

Between 2010 and 2013 three new fen raft spider populations were established in the Norfolk and Suffolk Broads. During this period almost six thousand test-tube reared spiderlings, together with thirty-two nurseries from the Pevensey Levels, potentially contributing upwards of another seventeen thousand tiny spiderlings, were released at the new sites.

The first two release sites were on the river Waveney grazing marshes. At Castle Marshes, and then at Calton Marshes two miles downstream, spiders were introduced over two successive years. Evidence from Redgrave & Lopham Fen and from captive rearing work suggested that the great majority of spiders breed when two years old and so introductions of spiderlings over two years were needed to emulate the age structure of natural populations. The first

Early morning dew on nursery web

introduction of test-tube spiderlings at Castle Marshes was made in early October 2010, only two weeks before the onset of a memorably harsh and prolonged winter. Spring 2011 was a quiet and anxious time. And then in late May they started to appear; smart, almost half grown and looking completely at home, basking and hunting amongst the well-armed protection of the water soldiers. In July adult females from the Pevensey Levels, released with their broods, augmented their numbers and went on to produce their own second broods on the ditches. Ghosts from the past, glowing balls of dew, lit up the water soldiers of the Waveney marshes. In 2012 the released spiderlings, bred if not born on the marshes, had their own nurseries; over forty were found during the summer, contributing many thousands of tiny spiderlings to the new population.

Encouraged by these early signs of success, the translocation programme moved north into the heart of the Broads. Although there can never be proof that fen raft spiders occurred here in the past, the linkages between the five river valleys that make up the Broads catchment, and the historical continuity of suitable habitat along them, make it improbable that this was anything other than the return of a native. In 2012 and 2013 spiders were released on the mid-reaches of the river Yare, to the east of Norwich. Again, the site seemed perfect - protected as a nature reserve, this time by the RSPB, with eight miles of grazing marshes managed for their rich and varied wetland wildlife. Like the lower Waveney marshes, but on a grander scale, many ditches in this extensive network were magnificent with water soldier – wide green highways along which the spider population could spread.

By 2013, nursery webs had become a common, almost familiar sight on the ditches at the Castle Marshes, the first introduction site. On one seventy metre stretch of ditch, where the water soldier was particularly dense, well over a hundred nurseries were built during the summer. One of the secrets of the spider's success, undetectable in the wild populations but revealed by these staged introductions, was that some individuals in the new populations were breeding after only one year. On the mid-Yare Marshes at least forty nurseries were recorded in the summer following their first introduction. First year breeding was found more commonly where spiderlings had been released in bottled nurseries in July. Although their early survival would have been much lower than amongst those cosseted in test tubes and released in autumn, for some of the survivors, the natural food supply was apparently so plentiful that they were able to grow and mature substantially faster. At the same time, amongst spiders reared in captivity, it was discovered that a small proportion could grow very slowly, delaying their maturity until their third year.

The early success of these new raft spider populations bodes well for their future in Broadland. Continuing assessment of the size and spread of the new populations, and of the maintenance of their genetic diversity, will help to inform future translocations within the Broads. If sustained, the success also offers the exciting possibility of establishing future populations without further need to take spiders from the natural populations. The new populations are already so dense in places that they themselves may soon be able to supply the adult females needed for further translocations.

Beyond the Broads

Although the spider's immediate future seems to lie in the Broads of Norfolk and Suffolk, the future of these wonderful and much-loved wetlands is in danger. Rising sea levels threaten them with salt water, forced ever further up the river systems by the embankments built to keep it out; Broadland is unlikely to remain a sanctuary for freshwater species in the long-term.

A spidery world: 6am on Castle Marshes

The search for new homes for Broadland wildlife, and particularly for species for which the Broads harbour internationally important populations, has already begun. It forms part of the rationale behind major fenland restoration projects now underway in the inland fen basin of East Anglia. This undraining of the fens, this undoing of nearly four centuries of human endeavour to tame and then obliterate a once-vast watery wilderness, is full of promise: the Great Fen project, reuniting the remnants of ancient Woodwalton and Holme Fens; the Wicken Vision, stretching from the last precious core of Wicken Fen back to the very doorstep of Cambridge; and Lakenheath Fen emerging triumphant from the former monotony of carrot fields next to the tiny fenland scrap of Botany Bay.

Already recognised as home by chattering masses of reed and sedge warblers and haunted by angular shadows of harriers, these fens are both new and ancient; their reed-swept horizons a bittern-booming echo of the past. Their resemblances to the old - the landscapes and the more mobile species – are restored with relative ease. But the return of former richness and rarity, and of the deep black peat on which both depend, are tasks not for years but for centuries. Fortunate among the many species lost, fen raft spiders may again be offered a helping hand to return as suitable new habitat develops. And, in this case, there is written evidence that this would be a true homecoming. The great naturalists of Nineteenth Century Cambridgeshire recorded *Dolomedes* at Wicken Fen and at other fens long lost to ditch and plough. Although some later arachnologists cast doubt on their records, meticulous recent literature research by arachnologist Ian Dawson reveals compelling accounts. The Rev. Leonard Jenyns, for example,[3] can hardly have been mistaken when he wrote that 'this fine species, sometimes measuring in the body an inch in length, is not uncommon in the fens'. Although these spiders were recorded as *fimbriatus* – British arachnologists of the time did not distinguish the two species - it is much more likely that this vast area of fen habitat would have supported *Dolomedes plantarius*. Jenyns noted too that *Dolomedes* occurred occasionally on ponds "as in the pond in the Kitchen garden at Wilbraham Temple". Records from artificial ponds are also typical of *plantarius*. On the continent it is recorded from monastic fish ponds and in the UK it was recorded by Evan Jones on the moat of Pevensey Castle in the early 1990s.

Away from East Anglia, the search for potential new homes for the fen raft spiders continues although large areas of overtly suitable habitat are in shorter supply. The predictions of climate change models suggest that the grazing marshes of the south coast may become unsuitable for the spiders over the next fifty years[4] and increase the pressure to look further north, beyond the triangle defined by the three natural populations.

Back at home

On Redgrave & Lopham Fen where the story began, translocation also had a role to play in the spider's recovery. Between 2010 and 2012, over two thousand home-grown test-tube spiderlings were released within two areas of the Fen from which they had been lost. With four pockets of population instead of two, and a programme of new pond creation to create better summer-wet corridors between them, the spiders finally had a better chance of repopulating a much greater area of the reserve. Monitoring the success of these introductions amongst the Fen's dense sedge beds is much more challenging than along the grazing marsh ditches downstream. Early indications though are that the spiders are starting to establish in the new areas, taking the first steps towards repopulation of their former range on the reserve.

Four-spotted
chaser

Meadow thistle - June

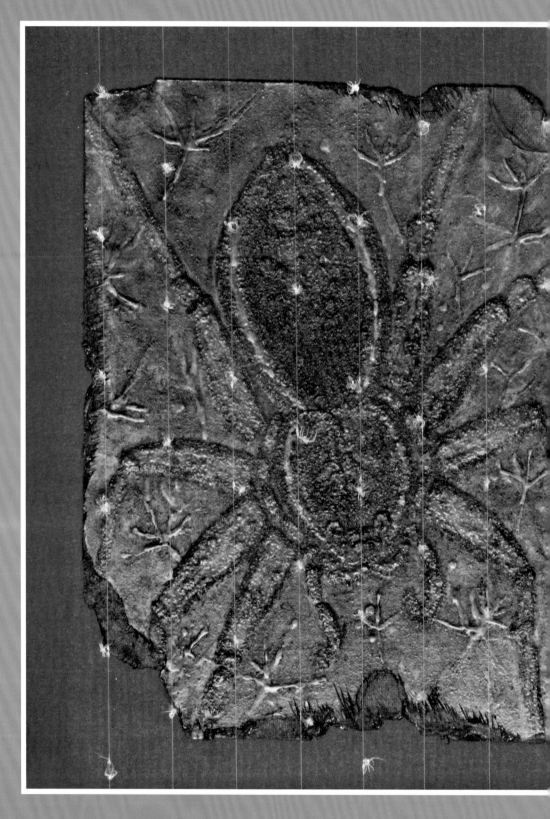

20 RELICS
Technique: Mixed medium
10" x 8.5" : 225mm x 215mm

A body of data about the Fen Raft Spider has been systematically compiled since the launch of the Species Recovery Programme. From the early 1990s onwards, skins found in the field have been collected and, subsequently, ways have been discovered for DNA to be successfully extracted. This has provided invaluable scientific evidence to monitor some of the changes that have happened to the population through this time.

This extensive research has created a multi-dimensional view of this species: reference to the past has helped to provide the best pathway forward in order to help the spider flourish.

This image represents *Dolomedes plantarius* as a species to illustrate the nature of reproduction in passing on a replica, visually intact although genetically changed, of what has gone before, generation following generation.

The viewer is invited to observe this image with the aid of a lens and engage with the intricate structure of the skins. They hang in suspended rows in front of a spider constructed as a collagraph plate using layers of carborundum grit and PVA. After it had been printed it was obvious that the plate, slightly raised in relief, suited the three dimensional aspect of the skins much better than the flat paper image taken from it, and so it remains as the representative of *Dolomedes plantarius* itself.

The tiny skins shed from spiderlings are too small to be seen in clear detail yet they hold a perfect ghostly impression of all the outward features of the spider, each a replica of its siblings as well as a replica of past generations extending through its ancestral line.

Spent spider skin

Larval casing of dragonfly

Dusk on the Fen

Beyond translocation

The time of writing is early days for the new fen raft spider populations but the story so far seems a simple one. The Redgrave & Lopham Fen spiders were indeed in need of a lift home. Down the road, down the Waveney, in ancestral haunts rescued from degradation and restored to former glories by conservation organisations, they show every sign of thriving. The first new population is already as dense on the water soldier ditches of Castle Marshes as on those of the Pevensey Levels. Perhaps the spider's slow response to improving conditions back at home on Redgrave & Lopham Fen was at least in part because, although adequate, this is sub-optimal habitat for them – their population there an outlier, a precarious and lone survivor of the changes that devastated so many wetlands, particularly in the post-war decades of the 20th Century.

If these early successes prove enduring, fen raft spiders may join the spectacular array of uncommon, beautiful and bizarre species to be enjoyed on sunny summer days in some of our richest wetland reserves. Perhaps in the process they will lose some of the fame that accompanies extreme rarity. But the threat of extinction is a tragic accolade - far better a future in which this species can act out its remarkable life history without the possibility of each generation being the last. Far better a species that can be enjoyed by the many not the few: one that will serve as a tribute to those who have fought to bring our damaged wetlands back to life; a torch bearer for sites and species still needing protection and a memorial to all those species more brown, less big, that fell, irretrievably and unnoticed, over the brink.

Threads

I have travelled a long way with fen raft spiders, literally as well as metaphorically. From the sunny coastal marshes of Sussex back home to the Norfolk-Suffolk borders, exhilarated by buzzards over the Downs, desperate in the sweltering queues at the Dartford tunnel and covering my precious cargo from unpredictable reactions at rushed service station stops. And then again, from my home at the source of the Waveney, with immense trepidation, to deliver these pioneers, now bottled with nurseries, or test-tube confined as spiderlings, to what I hoped would become their new homes downstream.

There have been journeys too with individual spiders, destined for stardom at public events, taking their own story to the wider world; on the Underground, in my bag, the spider is a secret talisman connecting me through darkness and human crush to the greater reality of sedge, mud and sunshine.

For many years my relationship with the spiders felt largely unshared and personal, best understood perhaps by the few who had previously studied their lives and lamented their decline. The prospect of the translocation programme ended any exclusivity. The research projects, so vital to changing the spider's fortunes, brought in students Phil Pearson and Marija Vugdelić at the University of East Anglia and Andrew Holmes at Nottingham University. The development of captive rearing techniques involved the expertise of Ian Bedford and his team at the John Innes Centre insectary in Norwich. Site assessment surveys across England enlisted the help of teams of expert arachnologists from the British Arachnological Society.

Once underway, the translocation programme was undoubtedly the most bizarre phase of the spider's story and perhaps, in retrospect, also of my life. Although successful captive rearing techniques were developed at the John Innes Centre insectary, the needed to mass produce spiders from Redgrave & Lopham Fen was, of financial necessity, largely a cottage industry.

My kitchen, close to their home on the Fen, became home to thousands of tiny spiderlings and their mothers. Sharing their secret lives, from the nocturnal spinning of egg sacs to the moulting contortions of tiny spiderlings, was an unforgettable privilege. Finding food for them – tiny flies, each to be delivered alive into a test tube without the spider inside escaping - and ensuring that their cotton wool bedding was free of infecting moulds, was a constant challenge. When the supply of fruit flies, captured in huge polythene bags as they rose in swarms from my compost bins, was depleted, the Konik ponies that graze the Fen came to the rescue with a copious supply of lesser dung flies. Searching the Fen, sweep net in hand, for fresh pony dung on which the minute flies sparkled and fizzed in incredible numbers, became a summer ritual. The grim times, though, were when the weather failed. The fly supply all-but dried up and, for my long-suffering displaced family, the necessity of eating *al fresco* suddenly lost its appeal.

In 2011 I was relieved to be helped in the task of spider rearing by three of the UK's Zoos, inspired and co-ordinated by Ian Hughes of BIAZA (the British & Irish Association of Zoos and Aquariums) and Dudley Zoo. A true warrior for wildlife, Ian's manic journeys, from his home in west Wales across the length and breadth of England, collecting and delivering test-tube spiderlings at precarious motorway service station rendezvous, became legendary. By 2012, another seven zoos had joined the team of spider foster parents, bringing with them skill, heart-warming enthusiasm and a wonderful opportunity to share the spider's story with a huge new audience. For many of these surrogate parents, the devotion of their months of caring became clear with the pain of letting go; the departure and release their spiderlings was accompanied by surreptitious tears.

Among the growing team of those now integral to delivering the new populations were the wardens of the new sites, both volunteers and staff. To help assure the spiders future we needed, and were fortunate to find, sites where their arrival would be not simply accommodated but embraced with enthusiasm. For me, the letting go, the passing on of ownership, turned out to be one of the unexpected rewards of the project. The optimism and pride in their new acquisitions, the excitement at the first sightings in spring, and the later arrival of egg sacs and then of nurseries, was a new dimension to the project. The spiders were in good hands.

Perhaps the most un-thought-of dimension at this turning point in the history of Redgrave & Lopham Fen's raft spiders was to have it reflected in art through Sheila Tilmouth's residency. I was fortunate to have had parents who immersed me in the natural world from birth and to have inherited just enough of my mother's artistic eye always to see beauty in natural form and detail. I was predisposed to the idea of using the visual arts to convey the story of these amazing and beautiful members of an often misunderstood and misrepresented group of animals.

As a fine artist, Sheila shared my love of detail which she explored with the analytical eye of a scientist. She wanted to understand and convey form in terms of function and attacked the arachnological literature with zeal. Her new love affair with the revelations of the microscope reminded me of my own unforgettable first encounters with the translucent world contained in a drop of pond water – *Volvox, Vorticella, Stentor, Euglena* - the litany of names as memorable as the revolving, flowing and waving creatures themselves. But as well as detail, Sheila also shared my love of form, big and bold but truthful in her prints. She captured the essence of this magnificent spider and distilled its complex stories into powerful images. Her images interpret – more succinctly and memorably than any words - the story of how this animal lives its life, of how close we came to losing it from Redgrave & Lopham Fen, and how this population is now being used to help secure a place for this, our grandest of spiders,[5] in Britain's future fauna.

Acknowledgements

This story owes everything to Eric Duffey. It may well have been a history, ending in the 1990s, had it not been for his vigilance for the fen raft spiders he discovered at Redgrave & Lopham Fen in 1956.

Since 1991 the Nature Conservancy Council and its successor organisations, English Nature and Natural England, have kept faith with, funded and led the Fen Raft Spider Species Recovery Programme. Natural England was joined, as funding partners, in the more recent translocation phase by the Broads Authority and the late and lamented BBC Wildlife Fund.

Ever since concern was first voiced about their decline in the 1970s, the Suffolk Wildlife Trust has led the conservation delivery for the spiders on its Redgrave & Lopham Fen reserve and, more recently, on the first, pioneering translocation sites. The Sussex Wildlife Trust has also been an integral project partner, supporting first of all research and survey work on its Pevensey Levels reserve and, more recently, the harvesting of female spiders from Pevensey to help stock the new East Anglia populations. The survey work and identification skills of the British Arachnological Society's volunteers continue to be an important element the project. Most recently, the RSPB has joined the partnership, supporting the establishment of the new population on its mid-Yare reserve in Broadland.

The contribution first of the John Innes Centre Insectary, and then of BIAZA zoos and collections, to the captive rearing of spiderlings transformed the viability of this element of the translocation work. The generosity of all of the zoos and collections in freely donating their time, facilities and expertise was an enormous contribution to the project. The foster parents included staff and volunteers at Beale Park, Bristol Zoological Gardens, Chessington World of Adventures, Chester Zoo, Dudley Zoological Gardens, Lakeland Wildlife Oasis, London Zoo, Reasheath College, The Deep at Hull and Tilgate Wildlife Centre. Dudley Zoo deserves particular mention for supporting much of Ian Hughes' work in co-ordinating the BIAZA work. Staff at the Institute of Zoology also gave their time and expertise to ensure the biosecurity of the BIAZA work. Geneticists Dr Sara Goodacre, and the late, great and much missed Professor Godfrey Hewitt, brought to the project new insights into the spider's past and the understanding needed to conserve their irreplaceable genetic diversity for the future. The sharp brains, fresh eyes and enormous volume of hard work of students Marija Vugdelić, Andrew Holmes, Phil Pearson (working with supervisor Bob James) and Alice Baillie underpinned the recent advances in conservation action for the spiders.

The Artist Residency was funded by grants from Arts Council England and the BBC Wildlife Fund, and supported by the Suffolk Wildlife Trust, British Arachnological Society and Buglife.

The support of organisations is delivered through the commitment of individuals, sadly too numerous to name individually. From the cheerful Suffolk Wildlife Trust volunteers who washed countless test tubes, through survey volunteers and site wardens and BIAZA foster parents, to the Natural England staff (both nationally and in the regions) who have fought the spider's corner for so many years, they have been collectively inspirational as well as vital to the project. The success of the new populations, and persistence of the old, is both tribute and reward to them all.

In addition to those mentioned in the text, and at reluctant risk of being invidious, amongst the many individuals to whom I am very grateful for help and support over two decades, I would particularly like to mention Wildlife Trust staff Arthur Rivett, Andrew Excell, Matt Gooch, Richard Young, Mike Harding and Dorothy Casey in Suffolk and Alice Parfitt in Sussex. Reserve volunteers Phil Lazaretti, George Batchelor and Jim Armes have been kindred spirits in sharing my own commitment to the project. In Sussex arachnologist Evan Jones' expert knowledge of the spider population on the Pevensey Levels has been invaluable and he and his wife Jane have generously helped with field work, morale and hospitality. Anna Jordan, working with Ian Bedford at the John Innes Centre Insectary, developed a memorable affinity with the spiders in her care and her meticulous observations advanced our understanding of the spiders' biology. At Natural England and its predecessor organisations, Martin Drake, Roger Key and David Heaver all gave consistent commitment to the project over many years, providing support and encouragement and finding funding from an ever-shrinking purse. More recently Andrea Kelly at the Broads Authority, and Tim Strudwick and Jane Sears at the RSPB, have been instrumental in helping to bring the translocation programme to the Norfolk Broads.

Special thanks to David Orr for allowing us to use his historic photographs of the Fen, to Arthur Rivett who scanned them for us and allowed us to use his photograph of a fen raft spider in a dipwell, and to John Harding of Francis Cupiss for use of his magnificent ancestral 1830's letter press.

Many friends and colleagues made invaluable comments on the text; James, Alice and Stephen Baillie, Rose Exall, Peter Smithers, Hannah Lawson, Ruth Bowen, Dorothy Casey, Mike Harding, David Heaver and the editorial team at Langford Press.

Finally we would like to thank Ian Langford for his support, for venturing to publish a book by two women about one spider, and for bending the remit of Langford Press from 'birds and people' to 'wildlife and people' to accommodate it.

Notes

Chapter 1

1.   Duffey, E. (1958) *Dolomedes plantarius* Clerk, a spider new to Britain, found in the Upper Waveney Valley. *Transactions of the Norfolk and Norwich Naturalists' Society*, **18**, 1-5.
2.   Platnik, N. (2008) *The world spider catalog*, version 14.0 http://research.amnh.org/iz/spiders/catalog/PISAURIDAE.html
3.   Kirby, P. (1990) *Dolomedes plantarius* in East Sussex. *Newsletter of the British Arachnological Society*, **58**, 8.
4.   Clark, M. (2004) Moving in – the fen raft spider in Wales. *Nature Cymru*, **10**, 38-40.
5.   World Conservation Monitoring Centre (1996) *Dolomedes plantarius*. In: the IUCN Red List of Threatened Species. Version 2014.2. <www.iucnredlist.org>. Downloaded on 06.08.2014.
6.   Bellamy, D.J. & Rose, F. (1960) The Waveney-Ouse valley fens of the Suffolk-Norfolk border. *Transactions of the Suffolk Naturalists' Society*, **11**, 367-385.
7.   Harding, M. (1993) Redgrave and Lopham Fen, East Anglia: a case study of change in flora and fauna due to ground water abstraction. *Biological Conservation*, **66**, 35-45.
8.   Duffey, E. (1991) The status of *Dolomedes plantarius* on Redgrave and Lopham Fen 1991. Unpublished report, English Nature, Peterborough.
9.   Bratton J. H. (ed)(1991) *British Red Data Book: 3. Invertebrates other than Insects*. JNCC, Peterborough.
10. Smith, H. (2000) The status and conservation of the fen raft spider (*Dolomedes plantarius*) at Lopham and Redgrave Fen NNR, UK. *Biological Conservation*, **95**, 153-164.

Chapter 2

1.   Gorb, S.N. & Barth, F.G. (1994) Locomotor behaviour during prey-capture of a fishing spider, *Dolomedes plantarius* (Araneae: Araneidae): galloping and stopping. *Journal of Arachnology*, 22, 89-93.
2.   Nyffeler, M. & Pusey, B.J. (2014) Fish Predation by Semi-Aquatic Spiders: A Global Pattern. PLoS ONE, **9** (6), e99459.
3.   Chitin is the main component of the cuticle of invertebrates including spiders and insects. Tough, protective and semi-transparent, it consists mainly of a nitrogen-containing polysaccharide.
4.   Suter, R.B. & Gruenwald, J. (2000) Predator avoidance on the water surface? Kinematics and efficacy of vertical jumping by *Dolomedes* (Araneae, Pisauridae). *Journal of Arachnology*, **28**, 201-210.
5.   Carico, J.E., (1973) The Nearctic Species of the Genus *Dolomedes* (Araneae: Pisauridae). *Bulletin of the Museum of Comparative Zoology*, **144**, 435-488.

Chapter 4

1. The UK Biodiversity Action Plan (UK BAP). http://jncc.defra.gov.uk/page-5155. Downloaded on 05.08.2014.
2. Any disturbance of this species or its habitat, including for its conservation, requires licencing by Natural England under the Wildlife and the Countryside Act, 1981.
3. Unpublished manuscript of *ca* 1870.
4. Leroy, B., Paschetta, M., Canard, A., Bakkenes, M., Isaia, M., & Ysnel, F. (2013) First assessment of effects of global change on threatened spiders: Potential impacts on *Dolomedes plantarius* (Clerck) and its conservation plans. *Biological Conservation*, **161**, 155–163.
5. Bristowe, W.S. (1958) *The World of Spiders*. Collins, London.

Further reading

Bristowe, W.S. (1958) *The World of Spiders*. Collins, London.

British Arachnological Society website – http://www.britishspiders.org.uk/

Dolomedes.org.uk – the fen raft spider website. http://www.dolomedes.org.uk

Foelix, R.F. (2011) *The biology of Spiders*. Oxford University Press, Oxford, UK.

Herberstein, M.E. (2011) *Spider Behaviour*. Cambridge University Press, Cambridge, UK.

Roberts, M.J. (1995) *Collins Field Guide - Spiders of Britain & Northern Europe*. Harper Collins, London, UK.

West, R.G. (2009) *From Brandon to Bungay. An exploration of the landscape history and geology of the Little Ouse and Waveney Rivers*. Suffolk Naturalists' Society, Ipswich, UK.

Other Titles in the Wildlife and People (W & P) Series.
(formerly B & P)

1 Eagle Days - Stuart Rae (B & P)

2 Red Kites - Ian Carter

3 Short-eared Owl - Don Scott

4 On the Rocks (sea birds) - Bryan Nelson and John Busby

5 Harriers and Honey Buzzards - Mike Henry

6 Derek Ratcliffe - Various authors

7 On The Margins - Helen Smith and Sheila Tilmouth

More to follow in the near future.